Math Warmups

BY
GARY COCHELL, Ed.D.

COPYRIGHT © 2001 Mark Twain Media, Inc.

ISBN 1-58037-168-X

Printing No. CD-1389

Mark Twain Media, Inc., Publishers
Distributed by Carson-Dellosa Publishing Company, Inc.

Table of Contents

Table of Contents

Introduction

This book is intended for the teacher who would like to start class with a quick mathematics warmup exercise or the parent who would like to challenge his or her student with a mathematics exercise that isn't just a worksheet problem from the daily routine. In either case, this book will furnish lots of good problems that the teacher/parent doesn't have to spend time coming up with. The target audience for this book is children in grades 5–8, but it certainly could be used by students in higher grades also.

The book is organized by a calendar that approximates the school calendar in many schools. There are four quarters of nine weeks each. Each week has ten warmup problems (two per day) that are divided into two distinct strands.

The first strand consists of five problems (one per day) of three different types: mental arithmetic, calculator usage, and estimation. On Monday and Friday, there are mental math problems. These problems are to be worked without pencil and paper or calculator. They are intended to sharpen one's mental capacity to do basic kinds of arithmetic. This skill is an integral part of doing mathematics with pencil and paper or calculator. On Tuesday and Thursday, there are calculator exercises. These exercises are intended to help the student learn appropriate calculator usage, and also to expand the student's knowledge of the mathematical functions on the calculator. Having access to a calculator is important in today's world, but knowing how to use it appropriately is critical. Wednesday is reserved for estimation problems. One often-overlooked mathematical skill today is the ability to estimate a calculation. This is a very practical skill that we benefit from in the real world. It is the skill that makes us more alert consumers.

The second strand consists of five problems (one per day) that are tied to a specific theme. The themes change from week to week, but some are repeated in different quarters. They are picked to cut across many different subject areas. In this way, I hope to interest the student who doesn't like mathematics as a subject in itself. But I make sure to find a clever way to bring a mathematical exercise into each theme.

Each problem in the book is presented in two different places. One place is intended for the eyes of the teacher/parent only. In this place, suggestions of pedagogy and the answer to the problem are presented. In a second, separate place, the problem (without answer) is presented in a "ticket" for the student. These tickets are grouped together so that the teacher can easily photocopy them for classroom usage.

I have picked the mathematical content of the problems to match some of the standards as set out in the National Council of Teachers of Mathematics' (NCTM) recent document, *Principles and Standards for School Mathematics* (or *Principles and Standards* for short.) This will be addressed more specifically in the next section of this book. I also picked the content so that it did not necessarily have to match what the teacher was doing in class. In fact, one of the benefits of warmup exercises in math class is that it allows the teacher the opportunity to return (even if only briefly) to these topics. In this way, there is more integration of mathematical content. The student can see that they keep using certain skills again and again.

I have several suggestions for the teacher about how they might use these warmup exercises. They are intended for the beginning of math class, but there is no reason why a teacher couldn't make use of them at other times in the math period. They might be used very effectively at the end when some students don't have any "regular" mathematics to work on. At the beginning of the period, I can see the teacher cutting out these photocopied tickets and giving the students a set three to five minutes to work on them. They can be worked on individually or in small groups. That is for the teacher to decide. After the students have finished, the teacher can easily collect the tickets

iv

(possibly using them for attendance purposes) and begin the daily lesson. However, sometimes the teacher might want to spend a few minutes going over some point made in the warmup exercise. How does the teacher get the student to give the warmups their full attention? They could be counted in the student's grade as a component of the homework grade, or they could be used to allow the student a daily bonus opportunity. In any event, these warmup exercises give the teacher a few minutes to take care of the bookkeeping chores that often accompany the beginning of class, while giving the students something to do besides just sitting there talking or waiting for class to begin. The students are actively involved in mathematics, which is hopefully interesting and probably a little different than what their daily lesson is going to be. For the teacher who has an extended time period for mathematics, having different and varied activities during the period is critical to a successful classroom. With all of this said, the main reason a teacher might want to use this book is to give the student an opportunity to mentally warm up and get ready for the day's math lesson without having to burden the teacher with thinking up his or her own special problems for that chore.

I also have several suggestions for the parent who wants to use this book at home. In today's world, it is the parent who sometimes has to take the added responsibility to ensure that his or her student achieves at the desired level. This book might give that kind of parent the material to supplement the student's daily mathematics work. Often as a parent, you don't really understand where your student is in a mathematics curriculum or how the teacher has intended for certain concepts and skills to be learned. So giving your student these kinds of supplementary exercises is often a safe and useful activity. Having your student work on one or two of these exercises each night before the next day's school would be a very good routine. The teacher suggestions in the book should be enough to help the parent understand the intent of the particular problem. The content of the exercises are such that the parent shouldn't have to be a teacher themselves to understand what is going on in the exercise. The parent can use the ticket much like a teacher in the classroom setting. They can either be cut out or photocopied and then given to the student to work on. In this way, the book can be used again for other members of the family.

—THE AUTHOR—

Challenge Problem   Name: _____

1. **States Geography Math:** Florida was the 27th state admitted _____
   to the United States. Its highest point is only 345 feet in elevation (Britton Hill). This ranks 50th among the states for their highest points. The state ranks third in water area (11,761 square miles). If this water area was a single circular lake, then how big would this lake be?

Florida

Friday   Name: _____

1. **Mental Math:** Perform the following division: $\frac{176}{32}$.   _____

2. **Guess That Number:** This number is a fraction.   _____
   Both the numerator and denominator are prime numbers, and the percent $66\frac{2}{3}\%$ can be represented by this fraction. Guess that number.

Monday   Name: _____

1. **Mental Math:** Calculate 17% of 300.   _____

2. **North American Animals:** A grey wolf or   _____
   timber wolf may weigh as much as 175 pounds and be over 6 feet long (including the tail). If 1 ounce = 29.5 grams, then how many kilograms (kg) does this timber wolf weigh?

v

Standards 2000

In 1989 the NCTM came out with *Curriculum and Evaluation Standards for School Mathematics.* The NCTM was the first professional organization to develop goals for teachers and the mathematics they teach in schools K-12. In the last 11 years these goals have had a profound effect on the mathematics taught in American schools. Many state departments of education have changed their mathematics curriculums to reflect the goals of the NCTM as put forth in the *Curriculum and Evaluation Standards for School Mathematics.* This document was not considered a onetime answer, but the beginning of an ongoing process for improving mathematics education in America. It was the intent of the original writers of the *Curriculum and Evaluation Standards for School Mathematics* that these goals be periodically reviewed, evaluated, and revised. Over the past five years, the NCTM has been involved in that process. The culmination of this was the publishing of the *Principles and Standards* in April 2000. Once again these revised goals for mathematics education should have an immediate and profound effect on what happens in many of America's mathematics classrooms. When you hear about **Standards** in reference to mathematics education in our schools, it will almost certainly be in reference to the new *Principles and Standards.*

I would like to take a few lines to tell you how these goals have influenced me in the writing of this book. It is important to look at a few words lifted directly from the preface of *Principles and Standards:*

> The recommendations in it are grounded in the belief that all students should learn important mathematical concepts and processes with understanding. *Principles and Standards* makes an argument for the importance of such understanding and describes ways students can attain it.

Clearly, learning important concepts and processes with understanding are the overriding goals I have in mind for this book.

Specifically, the strand 1 problems in this book consisting of mental calculation, calculator usage, and estimation are a result of my feelings about these processes and are supported by the *Principles and Standards:*

> Students should also develop and adapt procedures for mental calculation and computational estimation with fractions, decimals, and integers. Mental computation and estimation are also useful in many calculations involving percents. Because these methods often require flexibility in moving from one representation to another, they are useful in deepening students' understanding of rational numbers and helping them think flexibly about these numbers. (p. 220)

> The effective use of technology in the mathematics classroom depends on the teacher. Technology is not a panacea. As with any teaching tool, it can be used well or poorly. Teachers should use technology to enhance their students' learning opportunities by selecting or creating mathematical tasks that take advantage of what technology can do efficiently and well—graphing, visualizing, and computing. (p. 26)

Likewise, my inclusions of thematic story problems in the strand 2 exercises are intended so that "problem solving can and should be used to help students develop fluency with specific skills" (p. 52). Some of these skills include the ability to "express mathematical relationships using equations" (p. 394) as well as the ability to "develop an initial conceptual understanding of different uses of variables" (p. 395), and also the ability to "model and solve contextualized problems using various representations" (p. 395).

The purpose of this book is to put to practice many of these goals and help in some small way to make mathematics education better. I would urge you as teachers and/or parents to see that the important policy makers in your schools are aware of the *Principles and Standards.*

TEACHER PAGE: QUARTER 1: WEEK 1

The theme for week 1 is mathematics you might encounter at the mall. All of the #2 problems are based on this theme.

Monday

1. **Mental Math:** Add the following numbers: 27 + 39 + 18 + 21.
   (You might give students a hint about grouping certain "easy" pairs. Then the above simply becomes (27 + 18) + (39 + 21) = 45 + 60 = 105.)

2. **Mall Math:** If an item is on sale for 25% off the regular price and the regular price is $9.00, then what is the sale price?
   (You might want to clarify the difference between the mark-down and the sale price. The mark-down is $2.25, and thus the sale price is $9.00 - $2.25 = $6.75.)

Tuesday

1. **Calculator Math:** Add the following list of numbers:
   231 + 473 + 1029 + 2396 + 917 + 1483 + 798 + 1008.
   (You might have the students all start at a common signal and give them a time limit. Also, you might want to tell them the appropriate way to add a list of numbers with their calculators. The answer is 8335.)

2. **Mall Math:** If one in three got a free prize as they entered the cinema, then how many of the 72 customers got a prize?
   (You might want to mention proportions before the students start this problem. You then could set up the following proportion $\frac{1}{3} = \frac{?}{72}$, and solving for ?, you get 24 as the answer.)

Wednesday

1. **Estimation:** Estimate (to the nearest hundred) the following sum: 1672 + 734 + 189 + 2130.
   (You might give students a hint about rounding each number to the nearest hundred and then adding those numbers. You would get 1700 + 700 + 200 + 2100 = 4700.)

2. **Mall Math:** Which is a better buy: 64 oz. for $1.95 or 40 oz. for $1.49?
   (You might want to talk about the difference between dollars per ounce and ounces per dollar. When you calculate ounces per dollar, the first is 32.8205 ounces per dollar and the second is 26.8456 ounces per dollar. Therefore, the better buy is the first. You get more ounces for your dollar.)

TEACHER PAGE: QUARTER 1: WEEK 1

Thursday

1. **Calculator Math:** Perform the indicated operations: $\dfrac{21 \times 69 \times 117}{52 \times 38}$

   (You might have the students all start at a common signal and give them a time limit. Also you might want to tell them the appropriate way to multiply and divide with the calculator without using memory or writing down intermediate answers. With almost any calculator you would perform the following keystrokes: 21 x 69 = x 117 = / 52 = / 38 =. The answer is 85.7961.)

2. **Mall Math:** Look at the diagram below. If you start at one door at the mall and walk a complete "circuit," then how far did you walk?
   (You might want to tell your students that "circuit" here simply means the perimeter of the shape. Therefore the answer is 500 + 100 + 30 + 100 + 400 + 30 + 400 + 200 + 30 + 200 + 500 + 30 = 2520.)

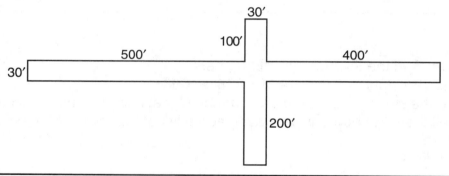

Friday

1. **Mental Math:** Multiply the following numbers: 5 x 78 x 2.
   (You might give students a hint about grouping certain "special" pairs to make the product easy. In this case, group the 5 x 2 to make the product above 78 x 10 = 780.)

2. **Mall Math:** Seven out of 63 stores at the mall are vacant. What percent of stores are in operation?
   (You might want to talk about the meaning of percent. If 7 out of 63 are vacant, then 56 out of 63 are in operation. So the percent is $\frac{56}{63}$ x 100 = 88.89%.)

Challenge Problem

1. **Mall Math:** What is the area of the floor in the mall pictured in the diagram above? (Problem two for Thursday.)
   (You might want to mention that the area can be broken into several simpler areas. One way is to break it into three rectangles: one 500′ by 30′, a second 330′ by 30′, and the third one 400′ by 30′. The total area would be (500 x 30) + (330 x 30) + (400 x 30) = 15,000 + 9,900 + 12,000 = 36,900 square feet.)

STUDENT PAGE: QUARTER 1: WEEK 1

Student

The theme for week 1 is mathematics you might encounter at the mall. All of the #2 problems are based on this theme.

Monday   Name: _____

1. **Mental Math:** Add the following numbers: 27 + 39 + 18 + 21.   _____

2. **Mall Math:** If an item is on sale for 25% off the regular price and the regular price is $9.00, then what is the sale price?   _____

Tuesday   Name: _____

1. **Calculator Math:** Add the following list of numbers: 231 + 473 + 1029 + 2396 + 917 + 1483 + 798 + 1008.   _____

2. **Mall Math:** If one in three got a free prize as they entered the cinema, then how many of the 72 customers got a prize?   _____

Wednesday   Name: _____

1. **Estimation:** Estimate (to the nearest hundred) the following sum: 1672 + 734 + 189 + 2130.   _____

2. **Mall Math:** Which is a better buy: 64 oz. for $1.95 or 40 oz. for $1.49?   _____

3

STUDENT PAGE: QUARTER 1: WEEK 1

Student

Thursday

Name: _____

1. **Calculator Math:** Perform the indicated operations:
   $$\frac{21 \times 69 \times 117}{52 \times 38}.$$

   _____

2. **Mall Math:** Look at the diagram below. If you start at one door at the mall and walk a complete "circuit," then how far did you walk?

   _____

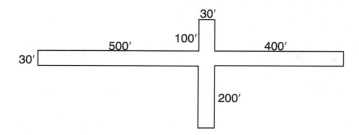

Friday

Name: _____

1. **Mental Math:** Multiply the following numbers: 5 x 78 x 2.

   _____

2. **Mall Math:** Seven out of 63 stores at the mall are vacant. What percent of stores are in opera-tion?

   _____

Challenge Problem

Name: _____

1. **Mall Math:** What is the area of the floor in the mall pictured in the diagram below?

   _____

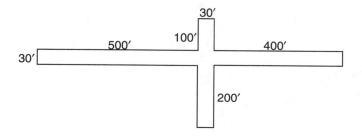

TEACHER PAGE: QUARTER 1: WEEK 2

The theme for week 2 is baseball. All of the #2 problems will be about some of the mathematics you might encounter in baseball.

Monday

1. **Mental Math:** Perform the following calculation: 54 - 29 + 31 - 21 + 39.
   (You might mention that sometimes you can group certain compatible numbers, e.g., in the above, group 31 with 39 and 29 with 21 to get the following: 54 + 70 - 50 = 4 + 70 = 74.)

2. **Baseball Math:** If Mark McGwire hits 20 home runs in his first 35 games, then at that rate how many would he hit in 140 games?
   (You might want to mention that proportions might be a good way of looking at this problem, i.e.,
   $\frac{20}{35} = \frac{?}{140}$ or $\frac{20 \times 4}{35 \times 4} = \frac{?}{140}$ and thus you get an answer of 80 home runs.)

Tuesday

1. **Calculator Math:** Perform the following calculation: [-2783 / (-75)] / (-17).
   (You might want to show your students how to use the +/- key or the (-) key on the calculator, and to talk about use of brackets for order of operations. With the (-) on many calculators you would perform the following keystrokes: (-) 2783 / (-) 75 = / (-) 17 =. The answer to this problem is -2.182745098.)

2. **Baseball Math:** If a baseball player has a batting average of .320, then how many hits did he get in his 400 at bats?
   (You might want to mention that a batting average is computed by dividing the number of hits by the number of at bats. Therefore, in the problem above we have $.320 = \frac{?}{400}$ and so the number of hits is 128.)

Wednesday

1. **Estimation:** Find a low and a high estimate of the following product: 261 x 117.
   (You might want to discuss rounding both of the numbers down to get a low estimate and rounding them both up to get a high estimate. In the above problem, that might be 250 x 100 = 25,000 for a low estimate and 300 x 150 = 45,000 for a high estimate.)

2. **Baseball Math:** A pitcher gave up one earned run in three and a half innings pitched. What was her earned run average?
   (You might want to mention that the earned run average of a pitcher is the number of runs a pitcher would give up in nine innings if they give them up at the same rate. That means you can solve this type of problem by solving the following proportion:
   $\frac{1}{3\frac{1}{3}} = \frac{?}{9}$. Therefore: $? = \frac{9}{3\frac{1}{3}} = \frac{9}{\frac{10}{3}} = \frac{27}{10} = 2.7$.)

5

TEACHER PAGE: QUARTER 1: WEEK 2

Thursday

1. **Calculator Math:** Perform the following multiplication: 1749200000000 x 0.000000000546.
   (You might want to show your students how to use scientific notation on their calculators. On many calculators you can use the EE or Exp key to enter numbers in scientific notation. Perform the following keystrokes: 1.7492 EE 12 x 5.46 EE (-) 10 =. The answer is 955.0632.)

2. **Baseball Math:** A major league player throws the baseball from third base to first base. How far does he have to throw the ball?
   (You need to mention that a baseball diamond is just a square with sides of 90 feet. Thus the distance from first base to third base is just the hypotenuse of an isosceles right triangle with a leg of 90 feet. This distance is $90\sqrt{2}$ feet or 127.279 feet.)

Friday

1. **Mental Math:** Multiply the following numbers: 2 x 96 x 25 x 2.
   (Here again you might mention that sometimes you can group certain compatible numbers in a multiplication problem, e.g., in the above, group the 2 with the 2 and 25 to get 96 x 100 = 9,600.)

2. **Baseball Math:** The pitcher has to throw the baseball 60.5 feet to home plate. If he throws it at the speed of 90 miles per hour (132 feet per second), then how long does it take the ball to reach home plate?
   (You might review the relationship Distance = Rate x Time. Using this relationship, the problem above can be solved by Time = 60.5/132 = 0.458333 seconds.)

Challenge Problem

1. **Travel Math:** You are making a trip to see your grandparents. It is a trip of 250 miles. The first 150 miles is easy driving and your parents will average 65 miles an hour on that stretch. You then stop 30 minutes for lunch and to get gas. The final 100 mile stretch is slower driving and your parents average only 50 miles per hour. How long does the trip to your grandparents' take?
   (You might make it clear that you need to think of it as the sum of three parts. Using Distance = Rate x Time, you can figure the first part of the trip by dividing Distance by Rate to get 150 / 65 = 2 hours 18 minutes. In the same way, you can figure the third part of the trip as 100 / 50 = 2 hours. Add the 30 minutes for the break, and you find that the trip takes 4 hours 48 minutes.)

STUDENT PAGE: QUARTER 1: WEEK 2

The theme for week 2 is baseball. All of the #2 problems will be about some of the mathematics you might encounter in baseball.

_____

Monday   Name:_____

1. **Mental Math:** Perform the following calculation: _____
   54 - 29 + 31 - 21 + 39.

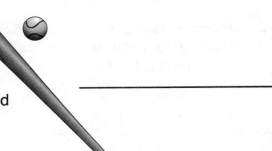

2. **Baseball Math:** If Mark McGwire hits 20 home runs   _____
   in his first 35 games, then at that rate, how many would
   he hit in 140 games?

_____

Tuesday   Name:_____

1. **Calculator Math:** Perform the following calculation: _____
   [-2783 / (-75)] / (-17).

2. **Baseball Math:** If a baseball player has a batting av-   _____
   erage of .320, then how many hits did he get in his 400 at
   bats?

_____

Wednesday   Name:_____

1. **Estimation:** Find a low and a high estimate of the following prod-   _____
   uct: 261 x 117.

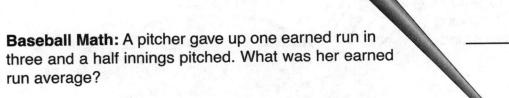

2. **Baseball Math:** A pitcher gave up one earned run in   _____
   three and a half innings pitched. What was her earned
   run average?

7

STUDENT PAGE: QUARTER 1: WEEK 2   **Student**

Thursday   **Name:** _____

1. **Calculator Math:** Perform the following multiplication: _____
    1749200000000 x 0.000000000546.

2. **Baseball Math:** A major league player throws the   _____
    baseball from third base to first base. How far does he
    have to throw the ball?

Friday   **Name:** _____

1. **Mental Math:** Multiply the following numbers: 2 x 96 x 25 x 2.   _____

2. **Baseball Math:** The pitcher has to throw the base-   _____
    ball 60.5 feet to home plate. If he throws it at the speed
    of 90 miles per hour (132 feet per second), then how
    long does it take the ball to reach home plate?

Challenge Problem   **Name:** _____

1. **Travel Math:** You are making a trip to see your grandparents. It   _____
    is a trip of 250 miles. The first 150 miles is easy driving and your
    parents will average 65 miles an hour on that stretch. You then
    stop 30 minutes for lunch and to get gas. The final 100 mile
    stretch is slower driving and
    your parents average only
    50 miles per hour. How long
    does the trip to your grand-
    parents' take?

TEACHER PAGE: QUARTER 1: WEEK 3

The theme for this week will be far out mathematics—the mathematics of astronomy. All of the #2 problems are based on this theme.

Monday

1. **Mental Math:** Perform the following multiplication: 27 x 102.
   (You might want to show your students how you can use the distributive property in some special cases to allow you to do the above problem mentally, i.e. 27 x 102 = 27 x (100 + 2) = 2700 + 54 = 2754.)

2. **Far Out Math:** If the moon is approximately 768,000 kilometers from Earth, how many miles is that? Remember that 5 kilometers is about 3.1 miles.
   (You might want to mention that a proportion could be used in this situation. Or you could mention "dimensional analysis" for this problem, i.e.,

$$768,000 \text{ km} \quad x \quad \frac{3.1 \text{ miles}}{5 \text{ km}} \quad = 476,160 \text{ miles.)}$$

Tuesday

1. **Calculator Math:** Perform the following calculation: [(23 + 71) x 13 - 456] / (29 + 88).
   (You might want to review order of operations with your students. You also might want to talk about using the parentheses key on the calculator. You might perform the keystrokes in the following way: ((23 + 71) x 13 - 456) / (29 + 88) =). The answer to the above problem is 6.547008547.)

2. **Far Out Math:** A light year is the distance that light travels in one year. If light travels at 186,282 miles per second, then how many miles are in a light year?
   (You might want to review Distance = Rate x Time. Then the above problem is solved by 186,282 miles per second x (365 x 24 x 60 x 60 seconds) = 5,874,589,152,000 miles.)

Wednesday

1. **Estimation:** Estimate the answer to the following division problem: 5589 / 72.
   (You might want to tell your students to look for compatible numbers in an estimation problem like the above. This one might be estimated by the following calculation: 5600 / 70 = 80.)

2. **Far Out Math:** The sun is approximately 93 million miles from Earth. How long does it take light to travel from the sun to the earth? Remember, light travels at 186,282 miles per second.
   (You might want to review Distance = Rate x Time. Thus Time in our problem = (93,000,000 / 186,282) seconds = 499.24 seconds.)

TEACHER PAGE: QUARTER 1: WEEK 3   Teacher

Thursday

1. **Calculator Math:** Perform the following calculation: $\dfrac{(1.47 \times 10^{12})\,(4.96 \times 10^{-15})}{3.62 \times 10^{-7}}$

   (You might want to review scientific notation and how to express numbers on the calculator in that notation. If your calculator has the EE or Exp key, then perform the following keystrokes: 1.47 EE 12 x 4.96 EE (-) 15 = / 3.62 EE (-) 7 =. The answer to this problem is 20141.43646.)

2. **Far Out Math:** The distance from the earth to the sun (approximately 93,000,000 miles) is called an Astronomical Unit (A.U.). Mercury is the closest planet to our sun. It is only 0.4 A.U. from the sun. How many miles is that?
   (You might want to talk about what a "unit" of measure represents. In this problem all you have to do is multiply 93,000,000 x 0.4. The answer is 37,200,000 miles.)

Friday

1. **Mental Math:** Perform the following division: 112 / 8.
   (You might want to tell your students that dividing the numerator and denominator by the same number keeps the answer the same, e.g., $\dfrac{112}{8} = \dfrac{56}{4} = \dfrac{28}{2} = 14$.)

2. **Far Out Math:** Pluto is approximately 3,673,500,000 miles from the sun. How many A.U.s is that?
   (You might want to look at problem #2 from Thursday for the definition of an A.U. Therefore, all you need to do is divide 3,673,500,000 by 93,000,000 to get the number of A.U.s. The answer is 39.5 A.U.s.)

Challenge Problems

1. **Mental Math:** Multiply the following numbers: 999,999 x 9.
   (You might think that this looks very hard to do mentally. But it really isn't all that difficult if you think of it as (1,000,000 - 1) x 9 or as one million 9's less one 9. In either case, you get simply 9,000,000 - 9 = 8,999,991.)

2. **Mental Math:** Multiply the following fractions: $\dfrac{3}{4} \times \dfrac{2}{9} \times \dfrac{6}{15}$.

   (In this problem, you might want to look at it as $\dfrac{3}{15} \times \dfrac{6}{9} \times \dfrac{2}{4} = \dfrac{1}{5} \times \dfrac{2}{3} \times \dfrac{1}{2} = \dfrac{2}{30} = \dfrac{1}{15}$.)

STUDENT PAGE: QUARTER 1: WEEK 3

The theme for this week will be far out mathematics—the mathematics of astronomy. All of the #2 problems are based on this theme.

Monday   Name: _____

1. **Mental Math:** Perform the following multiplication: 27 x 102.   _____

2. **Far Out Math:** If the moon is approximately 768,000 kilometers from Earth, how many miles is that? Remember that 5 kilometers is about 3.1 miles.   _____

Tuesday   Name: _____

1. **Calculator Math:** Perform the following calculation: _____
   [(23 + 71) x 13 - 456] / (29 + 88).

2. **Far Out Math:** A light year is the distance that light travels in one year. If light travels at 186,282 miles per second, then how many miles are in a light year?   _____

Wednesday   Name: _____

1. **Estimation:** Estimate the answer to the following division problem: 5589 / 72.   _____

2. **Far Out Math:** The sun is approximately 93 million miles from Earth. How long does it take light to travel from the sun to the earth? Remember, light travels at 186,282 miles per second.   _____

STUDENT PAGE: QUARTER 1: WEEK 3   Student

Thursday   Name: _____

1. **Calculator Math:** Perform the following calculation: _____
    $$\frac{(1.47 \times 10^{12})\,(4.96 \times 10^{-15})}{3.62 \times 10^{-7}}$$

2. **Far Out Math:** The distance from the earth to the sun (approxi-   _____
    mately 93,000,000 miles) is called an Astronomical Unit (A.U.).
    Mercury is the closest planet to our sun. It is only 0.4 A.U. from
    the sun. How many miles is that?

Friday   Name: _____

1. **Mental Math:** Perform the following division: 112 / 8.   _____

2. **Far Out Math:** Pluto is approximately 3,673,500,000 miles from   _____
    the sun. How many A.U.s is that?

Challenge Problems   Name: _____

1. **Mental Math:** Multiply the following numbers: 999,999 x 9.   _____

2. **Mental Math:** Multiply the following fractions: $\frac{3}{4} \times \frac{2}{9} \times \frac{6}{15}$.   _____

TEACHER PAGE: QUARTER 1: WEEK 4

The theme for week 4 is American history puzzles. All of the #2 problems for this week will be posed as puzzles about famous events in American history.

___

Monday

1. **Mental Math:** Perform the following multiplication: 32 x 25.
    (You might want to mention that in special cases you can multiply one of the multipliers by a certain number and divide the other multiplier by the same number to simplify the multiplication. The above would become 32 x 25 = 16 x 50 = 8 x 100 = 800.)

2. **American History Puzzle:** Missouri was granted statehood in what year? Hint! The sum of the digits is 12 and the tens digit is twice the units digit.
    (You might want to talk about problem-solving strategies. In particular, you might want to set up equations to help solve this problem. You can probably guess that the first two digits are 1 and 8. Therefore, $t + u = 3$ and $t = 2u$. With a little guess-and-check, you should find the answer, 1821.)

___

Tuesday

1. **Calculator Math:** Calculate the following: $\dfrac{\sqrt{3} - \sqrt{2}}{\sqrt{3} + \sqrt{2}}$

    (You might want to mention how to use the square root function and parentheses on the calculator. The calculation should be done without having to write down intermediate answers or using memory. If your calculator √ key works with the number "after," then perform the following keystrokes: (√ 3 - √ 2) / (√ 3 + √ 2 =). The answer is 0.101020514.)

2. **American History Puzzle:** The Constitution of the United States was created in what year? Hint! The number formed from the tens and units digits is 70 more than the number formed from the thousands and hundreds digits.
    (You might want to talk about problem-solving strategies. In particular, you might want to set up equations to help solve this problem. So you might set up $th + 70 = tu$ and then guess that the Constitution was written in the 1700s, so $th = 17$. Therefore, $tu = 87$, and the answer is 1787.)

___

Wednesday

1. **Estimation:** Estimate the area (in square centimeters) of your math book's front cover. Use a ruler if you have one.
    (You might want to review the area of a rectangle. Also you might want to discuss why, even if you use a ruler, the answer obtained this way is still an estimate.)

2. **American History Puzzle:** The American Civil War began in what year? Hint! The year is a prime number.
    (You might want to review what a prime number is and how to determine if a number is prime or composite. You should know that the answer is around 1860. However, 1860 is not a prime (it is divisible by 2), and so you might next try 1861, which is a prime. Since neither 1862, 1863, 1864, nor 1865 are primes, then a good guess for the answer is 1861.)

TEACHER PAGE: QUARTER 1: WEEK 4

Thursday

1. **Calculator Math:** Perform the following calculation: $1^2 + 2^2 + 3^2 + \ldots + 10^2$.
   (You might want to mention using the square function and doing the calculation with a single series of keystrokes. Perform the following keystrokes: $1\ x^2 + 2\ x^2 + 3\ x^2 + 4\ x^2 + 5\ x^2 + 6\ x^2 + 7\ x^2 + 8\ x^2 + 9\ x^2 + 10\ x^2 =$. The answer is 385.)

2. **American History Puzzle:** In what year did Martin Luther King, Jr., receive the Nobel Prize for Peace? Hint! This number is a composite with 491 as one of its factors.
   (You might want to review composite numbers at this time. You might try 491 x 2 = 982, 491 x 3 = 1473, and 491 x 4 = 1964. The year is 1964.)

Friday

1. **Mental Math:** Add $\frac{3}{8} + \frac{7}{12} + \frac{5}{8}$.

   (You might want to mention that even in addition of fractions to look for compatible numbers. In the above it is easy to add $\frac{3}{8}$ and $\frac{5}{8}$ to get 1. Thus the above becomes $1 + \frac{7}{12} = \frac{19}{12}$.)

2. **American History Puzzle:** Hawaii was granted statehood in what year? Hint! The sum of the digits is 24, and the units digit and the hundreds digit are the same.
   (You might want to talk about problem-solving strategies. In particular, you might want to set up equations to help solve this problem. In this problem, you might guess that the first two digits are 19. Then the units digit would have to be 9, and since the sum of the digits is 24, the tens digit would be 5. The answer is 1959.)

Challenge Problems

1. **American History Puzzle:** Samuel Morse developed the first practical telegraph instrument. The first telegraph message was sent on May 24 of the year whose digits sum to 17 and the units digit is equal to the tens digit. What is the year?
   (If you can guess that this happened in the 1800s, then the problem is relatively easy. Thus the sum of the units and tens digit would be 8, and if they are the same, then they must both be 4. The answer is 1844.)

2. **American History Puzzle:** In a famous court case concerning Brown vs. Board of Education of Topeka, Kansas, the Supreme Court ruled that segregation in schools was unconstitutional. This ruling was handed down on May 17 of the year that has 977 as a prime factor. What is the year?
   (You might need to review what a prime factor is. But when you do you will find it easy to figure out that the year is 977 x 2 = 1954.)

STUDENT PAGE: QUARTER 1: WEEK 4

Student

The theme for week 4 is American history puzzles. All of the #2 problems for this week will be posed as puzzles about famous events in American history.

Monday   Name: _____

1. **Mental Math:** Perform the following multiplication: 32 x 25.   _____

2. **American History Puzzle:** Missouri was granted statehood in   _____
   what year? Hint! The sum of the
   digits is 12 and the tens digit is twice
   the units digit.

Missouri

Tuesday   Name: _____

1. **Calculator Math:** Calculate the following: $\dfrac{\sqrt{3} - \sqrt{2}}{\sqrt{3} + \sqrt{2}}$ _____

2. **American History Puzzle:** The Constitution of the United States   _____
   was created in what year? Hint! The number formed from the
   tens and units digits is 70 more than the number formed from
   the thousands and hundreds digits.

Wednesday   Name: _____

1. **Estimation:** Estimate the area (in square centimeters) of your   _____
   math book's front cover. Use a ruler if you have one.

2. **American History Puzzle:** The   _____
   American Civil War began in
   what year? Hint! The year is a
   prime number.

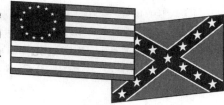

15

STUDENT PAGE: QUARTER 1: WEEK 4   Student

Thursday   **Name:** _____

1. **Calculator Math:** Perform the following calculation: _____
   $1^2 + 2^2 + 3^2 + \ldots + 10^2$.

2. **American History Puzzle:** In what year
   did Martin Luther King, Jr., receive the   _____
   Nobel Prize for Peace? Hint! This num-
   ber is a composite with 491 as one of its
   factors.

Friday   **Name:** _____

1. **Mental Math:** Add: $\dfrac{3}{8} + \dfrac{7}{12} + \dfrac{5}{8}$.   _____

2. **American History Puzzle:** Hawaii was granted statehood in what   _____
   year? Hint! The sum of the dig-
   its is 24, and the units digit and
   the hundreds digit are the
   same.

Hawaii

Challenge Problems   **Name:** _____

1. **American History Puzzle:** Samuel Morse developed the first   _____
   practical telegraph instrument. The first tele-
   graph message was sent on May 24 of the
   year whose digits sum to 17 and the units
   digit is equal to the tens digit. What is
   the year?

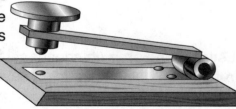

2. **American History Puzzle:** In a famous court case concerning   _____
   Brown vs. Board of Education of Topeka, Kansas, the Supreme
   Court ruled that segregation in schools was unconstitutional. This
   ruling was handed down on May 17 of the year that has 977 as
   a prime factor. What is the year?

16

TEACHER PAGE: QUARTER 1: WEEK 5

The theme for week 5 is famous Greek mathematicians. All of the #2 problems are based on this theme.

Monday

1. **Mental Math:** Perform the following multiplication: 54 x 11.
   (You might want to mention using the distributive property as a possible way of looking at this multiplication mentally. Then the problem can be dealt with in the following manner:
   54 x (10 + 1) = 54 x 10 + 54 x 1 = 540 + 54 = 594.)

2. **Greek Mathematicians:** Pythagoras was a Greek mathematician who lived in the sixth century B.C.E. You probably recognize the name from the theorem about right triangles that bears his name. Pythagoras called 6 a perfect number because it is the sum of its proper divisors, 1 + 2 + 3 = 6. There is one other perfect number less than 100. Find it.
   (You might want to mention what a proper divisor is. With a little work you will find that 1 + 2 + 4 + 7 + 14 = 28, and so the perfect number is 28.)

Tuesday

1. **Calculator Math:** Calculate the following: $2^2 + 2^3 + 2^4 + 2^5 + 2^6 + 2^7$.
   (You might want to show your students how to use their ^ keys. Perform the following keystrokes: 2 ^ 2 = + 2 ^ 3 = + 2 ^ 4 = + 2 ^ 5 = + 2 ^ 6 = + 2 ^ 7 =. The answer is 252.)

2. **Greek Mathematicians:** Euclid was a Greek mathematician who lived around 300 B.C.E. Most of the geometry that you study in school is called Euclidean geometry. His book, *Elements,* contained the mathematics that was deemed necessary for all "educated" people of this time. In *Elements* Euclid showed a way to easily find the greatest common divisor of a pair of whole numbers. See if you can find the greatest common divisor of 72 and 120.
   (You might want to mention what is meant by greatest common divisor. In this problem, you can just list the divisors of each number and then pick out the largest common one in the lists. The answer here is 24.)

Wednesday

1. **Estimation:** Estimate the following: $\dfrac{4.79 \times 21.03}{9.67}$
   (You might want to mention rounding these numbers to compatible numbers: $\dfrac{5 \times 20}{10} = 10$.)

2. **Greek Mathematicians:** Archimedes was a Greek mathematician who lived around 250 B.C.E. in the city of Syracuse on the island of Sicily. Besides being a mathematician, he invented many devices incorporating levers and pulleys. Some say that he discovered the following formula to find the area of a triangle $A = \sqrt{s(s-a)(s-b)(s-c)}$ where a, b, and c are the sides of the triangle and s is the semi-perimeter of the triangle. Find the area of a triangle whose sides are 4, 7, and 9.
   (You might want to clarify what is meant by semi-perimeter. In this problem, the semi-perimeter is $s = (4 + 7 + 9) / 2 = 10$, and so $A = \sqrt{10 \times 6 \times 3 \times 1}$. The answer here is $\sqrt{180}$.)

TEACHER PAGE: QUARTER 1: WEEK 5

Thursday

1. **Calculator Math:** Perform the following calculation:

$$\sqrt{\sqrt{\sqrt{\sqrt{2}+1}+1}+1}.$$

(You might want to mention to your students that they need to start at the inside and work out. If your calculator has the ANS key, then perform the following keystrokes: √ 2 + 1 = √ ANS + 1 = √ ANS + 1 =. The answer is 1.611847754.)

2. **Greek Mathematicians:** Eratosthenes was a Greek mathematician who lived around 200 B.C.E. in the famous city of Alexandria. He was a man who had many talents besides mathematics. He created an early map of the world as he knew it at that time. One of his mathematical inventions, a sieve, was an easy way to find prime numbers. Can you find all the prime numbers between 2 and 50?
(You might want to mention what a prime number is. The prime numbers are 2, 3, 5, 7, 11, 13, 17, 19, 23, 29, 31, 37, 41, 43, and 47.)

Friday

1. **Mental Math:** Perform the following subtraction of mixed numerals: $8\frac{2}{9} - 3\frac{7}{9}$.

(You might want to mention that if you add the same quantity to both numbers, then the answer remains the same. In this problem, you might add $\frac{2}{9}$ to both quantities giving $8\frac{4}{9} - 4 = 4\frac{4}{9}$.)

2. **Greek Mathematicians:** Diophantus is believed to be a Greek mathematician of around 250 C.E. who probably lived and worked in Alexandria. He is one of the "fathers" of the subject of algebra. See if you can solve one of his easier problems: Find two numbers whose sum is 20 and whose product is 96.
(You might want to mention problem-solving strategies. In particular, you might want to set up equations to help solve this problem. You could set up $a + b = 20$ and $a \times b = 96$. With a little guess-and-check, you can find the numbers are 8 and 12.)

Challenge Problem

1. **History of Math:** In ancient Mesopotamia (today this is the country of Iraq), a base 60 positional number system was developed. It was very advanced for its time. In a base 60 positional number system, the place values are units, sixties, sixties-squared, and so on, instead of units, tens, hundreds (tens-squared), and so on as in our base 10 positional system. For example, the base 60 number 1, 20, 3 would represent 3 ones plus 20 sixties plus 1 sixty-squared, which would be 3 x 1 + 20 x 60 + 1 x 3600 = 3 + 1200 + 3600 = 4802. What does the base 60 number 3, 15, 30 represent?
(The answer would be 30 ones plus 15 sixties plus 3 sixty-squared or 30 x 1 + 15 x 60 + 3 x 3600 = 30 + 900 + 10800 = 11730.)

STUDENT PAGE: QUARTER 1: WEEK 5

The theme for week 5 is famous Greek mathematicians. All of the #2 problems are based on this theme.

Monday   Name: _____

1. **Mental Math:** Perform the following multiplication: 54 x 11.

2. **Greek Mathematicians:** Pythagoras was a Greek mathematician who lived in the sixth century B.C.E. You probably recognize the name from the theorem about right triangles that bears his name. Pythagoras called 6 a perfect number because it is the sum of its proper divisors, 1 + 2 + 3 = 6. There is one other perfect number less than 100. Find it.

Tuesday   Name: _____

1. **Calculator Math:** Calculate the following:
   $2^2 + 2^3 + 2^4 + 2^5 + 2^6 + 2^7$.

2. **Greek Mathematicians:** Euclid was a Greek mathematician who lived around 300 B.C.E. Most of the geometry that you study in school is called Euclidean geometry. His book, *Elements,* contained the mathematics that was deemed necessary for all "educated" people of this time. In *Elements* Euclid showed a way to easily find the greatest common divisor of a pair of whole numbers. See if you can find the greatest common divisor of 72 and 120.

Wednesday   Name: _____

1. **Estimation:** Estimate the following: $\dfrac{4.79 \times 21.03}{9.67}$

2. **Greek Mathematicians:** Archimedes was a Greek mathematician who lived around 250 B.C.E. in the city of Syracuse on the island of Sicily. Besides being a mathematician, he invented many devices incorporating levers and pulleys. Some say that he discovered the following formula to find the area of a triangle $A = \sqrt{s(s-a)(s-b)(s-c)}$ where a, b, and c are the sides of the triangle and s is the semi-perimeter of the triangle. Find the area of a triangle whose sides are 4, 7, and 9.

19

STUDENT PAGE: QUARTER 1: WEEK 5   Student

Thursday   Name: _____

1. **Calculator Math:** Perform the following calculation: _____

$$\sqrt{\sqrt{\sqrt{\sqrt{2}} + 1} + 1} + 1$$

2. **Greek Mathematicians:** Eratosthenes was a Greek mathema-   _____
   tician who lived around 200 B.C.E. in the famous city of Alexandria. He was a man who had many talents besides mathematics. He created an early map of the world as he knew it at that
   time. One of his mathematical inventions, a sieve, was an easy
   way to find prime numbers. Can you find all the prime numbers
   between 2 and 50?

Friday   Name: _____

1. **Mental Math:** Perform the following subtraction of mixed nu-   _____
   merals: $8\frac{2}{9} - 3\frac{7}{9}$.

2. **Greek Mathematicians:** Diophantus is believed to be a Greek   _____
   mathematician of around 250 C.E. who probably lived and worked
   in Alexandria. He is one of the "fathers" of the subject of algebra.
   See if you can solve one of his easier problems: Find two numbers whose sum is 20 and whose product is 96.

Challenge Problem   Name: _____

1. **History of Math:** In ancient Mesopotamia (today this is the coun-   _____
   try of Iraq), a base 60 positional number system was developed.
   It was very advanced for its time. In a base 60 positional number system, the place values are units, sixties, sixties-squared,
   and so on, instead of units, tens, hundreds (tens-squared),
   and so on as in our base 10 positional system. For example,
   the base 60 number 1, 20, 3 would represent 3 ones plus 20
   sixties plus 1 sixty-squared, which would be 3 x 1 + 20 x 60 +
   1 x 3600 = 3 + 1200 + 3600 = 4802. What does the base 60
   number 3, 15, 30 represent?

TEACHER PAGE: QUARTER 1: WEEK 6

The theme for week 6 is States Geography Math, the mathematics that will help you learn some interesting facts about the geography of our states. All of the #2 problems are based on this theme.

Monday

1. **Mental Math:** Perform the following division: 2103 / 25.
   (You might want to mention that if you multiply both the numerator and denominator by the same quantity, then the answer remains the same. In this problem, multiply both the numerator and denominator by 4 to get 8412 / 100 = 84.12.)

2. **States Geography Math:** Alaska is the biggest state in land area (615,230 square miles). If Alaska were a square, then how big would the sides be?
   (You might want to review what area means and how to find the area of a square. Thus $(side)^2$ = 615,230 and taking the square root, you get the length of each side. The answer is a square of sides 784 miles.)

Tuesday

1. **Calculator Math:** Calculate $(2.5)^6$ without using your ^ key (if you have one). Also try it with the ^ key if you have it.
   (You might want to review raising a number to a power. Then doing this calculation without the ^ key would require you to perform 2.5 x 2.5 x 2.5 x 2.5 x 2.5 x 2.5 =. You should get an answer of 244.140625.)

2. **States Geography Math:** Colorado has many mountains that go over 14,000 feet in elevation. Its highest peak, Mount Elbert, is at an elevation whose square root is 120.137. What is its elevation?
   (You might want to review what square and square root mean. Thus you have $\sqrt{elevation}$ = 120.137, and you need to square both to get *elevation* = 14,433 feet.)

Wednesday

1. **Estimation:** Use estimation skills to place the decimal point in the following multiplication:
   54.387 x 0.135 = 73422.
   (You might want to tell your students to think of the problem as approximately $\frac{1}{10}$ of 50, which is equal to 5. Thus the decimal point would fit between the 7 and 3.)

2. **States Geography Math:** Texas is the second-largest state in land area (267,277 square miles). Texas is what percent of Alaska's land area (615,230 square miles)?
   (You might want to review the meaning of percent. Thus you would have to set up

   $\frac{267,277}{615,230}$ x 100 to get the percentage. The answer here is 43.44%.)

21

TEACHER PAGE: QUARTER 1: WEEK 6

Thursday

1. **Calculator Math:** Multiply 723.81 x 0.0375 without using the decimal point key on the calculator.
    (You might want to review the way we do multiplication by hand and how that might help students determine where to place the decimal point. Perform 72381 x 375 =. You will get 27142875, and then you will count back six places for the decimal point. The answer is 27.142875.)

2. **States Geography Math:** Delaware is the 49th state in land area (2,397 square miles). It only has three counties. If the counties were equal in area and square in shape, how big would each square be?
    (You might want to review the meaning of area and the formula for the area of a square. Each county would have an area of 799 square miles. Then take the square root of 799 to get the length of each side. The answer here is 28.3 miles on a side.)

Friday

1. **Mental Math:** Multiply: $\frac{7}{16}$ x 64.

    (You might mention that you can think of this problem as $7 \times \frac{1}{16} \times 64$ and then you can do the $\frac{1}{16}$ x 64 first and get 4. Thus the answer is 7 x 4 = 28.)

2. **States Geography Math:** Wyoming is the ninth-largest state in land area (97,819 square miles). Its lowest point (Belle Fourche River) in elevation is higher than many states' highest points. The sum of the digits in this four-digit number is 21, and the thousands digit divides both the tens and units digits (they are the same). The hundreds digit is 0. What is the elevation of Belle Fourche River?
    (You might want to review problem-solving strategies and in particular you might want to review setting up equations. Since the hundreds digits is 0, we would set up $th + te + u = 21$. With a little guess-and-check using the additional fact that th must divide both te and u evenly, you get $th = 3$, $te = 9$, and $u = 9$. The answer here is 3,099 feet.)

Challenge Problem

1. **States Geography Math:** Montana is the fourth-largest state in land area (145,556 square miles) and it has only 56 counties. If all the counties were the same size and each was a square in shape, then what would the dimensions of each county be?
    (You might want to review the area formula for a square. First you need to divide the 145,556 square miles by 56 to get 2,599 square miles per county. Then taking the square root of 2,599, you get a square of dimension approximately 51 miles.)

STUDENT PAGE: QUARTER 1: WEEK 6 **Student**

The theme for week 6 is States Geography Math, the mathematics that will help you learn some interesting facts about the geography of our states. All of the #2 problems are based on this theme.

Monday   Name:_____

1. **Mental Math:** Perform the following division: 2103 / 25.   _____

2. **States Geography Math:** Alaska is the biggest state in land area (615,230 square miles). If Alaska were a square, then how big would the sides be?   _____

Tuesday   Name:_____

1. **Calculator Math:** Calculate $(2.5)^6$ without using your ^ key (if you have one). Also try it with the ^ key if you have it.   _____

2. **States Geography Math:** Colorado has many mountains that go over 14,000 feet in elevation. Its highest peak, Mount Elbert, is at an elevation whose square root is 120.137. What is its elevation?   _____

Wednesday   Name:_____

1. **Estimation:** Use estimation skills to place the decimal point in the following multiplication: 54.387 x 0.135 = 73422.   _____

2. **States Geography Math:** Texas is the second-largest state in land area (267,277 square miles). Texas is what percent of Alaska's land area (615,230 square miles)?   _____

STUDENT PAGE: QUARTER 1: WEEK 6   Student

Thursday   Name: _____

1. **Calculator Math:** Multiply 723.81 x 0.0375 without using the _____
   decimal point key on the calculator.

2. **States Geography Math:** Delaware is the 49th state in land _____
   area (2,397 square miles). It only has
   three counties. If the counties were
   equal in area and square in shape,
   how big would each square be?

DECEMBER 7, 1787

Delaware

Friday   Name: _____

1. **Mental Math:** Multiply: $\frac{7}{16}$ x 64.   _____

2. **States Geography Math:** Wyoming is the ninth-largest state in _____
   land area (97,819 square miles). Its lowest point (Belle Fourche
   River) in elevation is higher than
   many states' highest points. The sum
   of the digits in this four-digit number
   is 21, and the thousands digit divides
   both the tens and units digits (they
   are the same). The hundreds digit is
   0. What is the elevation of Belle
   Fourche River?

Wyoming

Challenge Problem   Name: _____

1. **States Geography Math:** Montana is the fourth-largest state in _____
   land area (145,556 square miles) and
   it has only 56 counties. If all the counties were the same size and each was
   a square in shape, then what would
   the dimensions of each county be?

MONTANA

Montana

TEACHER PAGE: QUARTER 1: WEEK 7

The theme for week 7 will be making change. You will be given the amount of a purchase and the amount of money given to pay for the purchase. You are to figure out the change to give back using the least amount of bills and coins in your cash register. The register contains the usual coins (pennies, nickels, dimes, and quarters) and the usual bills (ones, fives, tens, and twenties). All of the #2 problems are based on this theme.

Monday

1. **Mental Math:** Perform the following subtraction: 284 - 38.
    (You might mention that you can add the same number to both numbers in a subtraction problem and the answer is the same. So in this problem if we add 2 to both numbers, we get 286 - 40 = 246.)

2. **Make Change:** The purchase is $0.59, and the amount tendered is $1.00.
    (You might suggest that your students try making change mentally by adding up to the amount tendered. Then they could check it with pencil-and-paper methods. The answer is 1-P, 1-N, 1-D, and 1-Q.)

Tuesday

1. **Calculator Math:** Calculate the following:
$$\cfrac{1}{1 + \cfrac{1}{1 + \cfrac{1}{4}}}$$

    (You might want to show your students how they could use the reciprocal key on their calculator to do a problem like this quickly and easily without storing, using parentheses, or writing down intermediate answers. You can perform the following keystrokes: $4\ x^{-1} + 1 = x^{-1} + 1 = x^{-1}$. The answer is 0.555555555 or $\frac{5}{9}$.)

2. **Make Change:** The purchase is $1.36, and the amount tendered is $5.00.
    (You might suggest that your students try making change mentally by adding up to the amount tendered. Then they could check it with pencil-and-paper methods. The answer is 4-P, 1-D, 2-Q, and 3-$1.)

Wednesday

1. **Estimation:** Estimate 41.3% of $123.74.
    (You might want to mention that one way of doing this is to think in terms of fractions. 41.3% is approximately 40%, which is $\frac{2}{5}$. Hence you are wanting to find $\frac{2}{5}$ of $125, which would be $50.)

2. **Make Change:** The purchase is $6.17, and the amount tendered is $20.00.
    (You might suggest that your students try making change mentally by adding up to the amount tendered. Then they could check it with pencil-and-paper methods. The answer is 3-P, 1-N, 3-Q, 3-$1, and 1-$10.)

TEACHER PAGE: QUARTER 1: WEEK 7

Thursday

1. **Calculator Math:** Calculate 41.3% of $123.74 without using the percent key on your calculator. (You might review converting percents to their decimal form. Then it is simply a multiplication problem 0.413 x 123.74, and the answer is $51.10.)

2. **Make Change:** The purchase is $1.38, and the amount tendered is $2.03. (You might suggest that your students try making change mentally by adding up to the amount tendered. Then they could check it with pencil-and-paper methods. The answer is 1-N, 1-D, and 2-Q.)

Friday

1. **Mental Math:** The plane was scheduled to leave at 10:15 a.m., but actually left at 11:50 a.m. How many minutes was the plane late? (You might want to review clock arithmetic. Here you take 15 from 50 to get 35 minutes and 10 from 11 to get 1 hour. Therefore the answer is 95 minutes.)

2. **Make Change:** The purchase is $47.83, and the amount tendered is $100.00. (You might suggest that your students try making change mentally by adding up to the amount tendered. Then they could check it with pencil-and-paper methods. The answer is 2-P, 1-N, 1-D, 2-$1, 1-$10, and 2-$20.)

Challenge Problems

1. **Make Change:** The purchase is $53.11, and the amount tendered is $60.01. (You might suggest that your students try making change mentally by adding up to the amount tendered. Then they could check it with pencil-and-paper methods. The answer is 1-N, 1-D, 3-Q, 1-$1, and 1-$5.)

2. **Make Change:** The purchase is $72.53, and the amount tendered is $100.03. (You might suggest that your students try making change mentally by adding up to the amount tendered. Then they could check it with pencil-and-paper methods. The answer is 2-Q, 2-$1, 1-$5, and 1-$20.)

3. **Make Change:** The purchase is $32.31, and the amount tendered is $40.06. (You might suggest that your students try making change mentally by adding up to the amount tendered. Then they could check it with pencil-and-paper methods. The answer is 3-Q, 2-$1, and 1-$5.)

4. **Make Change:** The purchase is $17.63, and the amount tendered is $50.13. (You might suggest that your students try making change mentally by adding up to the amount tendered. Then they could check it with pencil-and-paper methods. The answer is 2-Q, 2-$1, 1-$10, and 1-$20.)

STUDENT PAGE: QUARTER 1: WEEK 7 | **Student**

The theme for week 7 will be making change. You will be given the amount of a purchase and the amount of money given to pay for the purchase. You are to figure out the change to give back using the least amount of bills and coins in your cash register. The register contains the usual coins (pennies, nickels, dimes, and quarters) and the usual bills (ones, fives, tens, and twenties). All of the #2 problems are based on this theme.

Monday   Name: _____

1. **Mental Math:** Perform the following subtraction: 284 - 38.   _____

2. **Make Change:** The purchase is $0.59, and the amount tendered is $1.00.   _____

Tuesday   Name: _____

1. **Calculator Math:** Calculate the following:
$$\dfrac{1}{1+\dfrac{1}{1+\dfrac{1}{4}}}$$
_____

2. **Make Change:** The purchase is $1.36, and the amount tendered is $5.00.   _____

Wednesday   Name: _____

1. **Estimation:** Estimate 41.3% of $123.74.   _____

2. **Make Change:** The purchase is $6.17, and the amount tendered is $20.00.   _____

27

STUDENT PAGE: QUARTER 1: WEEK 7   **Student**

Thursday   Name: _____

1. **Calculator Math:** Calculate 41.3% of $123.74 without using the   _____
   percent key on your calculator.

2. **Make Change:** The purchase is $1.38, and the   _____
   amount tendered is $2.03.

Friday   Name: _____

1. **Mental Math:** The plane was scheduled to leave at 10:15 a.m.,   _____
   but actually left at 11:50 a.m. How many minutes was the plane
   late?

2. **Make Change:** The purchase is $47.83, and the   _____
   amount tendered is $100.00.

Challenge Problems   Name: _____

1. **Make Change:** The purchase is $53.11, and the amount ten-   _____
   dered is $60.01.

2. **Make Change:** The purchase is $72.53, and the amount ten-   _____
   dered is $100.03.

3. **Make Change:** The purchase is $32.31, and the amount ten-   _____
   dered is $40.06.

4. **Make Change:** The purchase is $17.63, and the   _____
   amount tendered is $50.13.

TEACHER PAGE: QUARTER 1: WEEK 8

The theme for week 8 is clock geometry. You will be given a time (assume the hour hand is always right on the number), and you will have to figure out the angle from the big hand to the small hand plus the distance along the circumference from the big hand to the small hand. The radius of the clock is 1 unit. All of the #2 problems will be based on this theme.

Monday

1. **Mental Math:** Add the following numbers: 43 + 32 + 37 + 18.
   (You might mention that sometimes you encounter compatible numbers in an addition problem. If so, then add those first. In this case we have 43 + 37 = 80 and 32 + 18 = 50. Hence the answer is 80 + 50 = 130.)

2. **Clock Geometry:** The time is 10:45. What is the angle, and what is the distance?
   (You might want to review time on a non-digital clock and use the fact that there are 30 degrees between each hour on the clock face. Also, the circumference of this circle of radius 1 is Diameter x π = 2 x π = 6.28…. Therefore, between the small hand and the big hand for 10:45 would be 30 degrees, and hence the distance on the circumference would be 30/360 of the 6.28 total circumference, i.e., approximately 0.52$\overline{3}$.)

Tuesday

1. **Calculator Math:** Calculate the following: $\dfrac{[23 \times (17 + 59) - 147] \times (74 - 36)}{(173 - 52)}$
   (You might want to talk about order of operations and the use of parentheses on the calculator. Also you should stress that the calculation can be done without using memory or writing down intermediate answers. You can perform the following keystrokes (23 x (17 + 59) - 147) x (74 - 36) / (173 - 52) =. The answer is 502.793.)

2. **Clock Geometry:** The time is 7:55. What is the angle, and what is the distance?
   (You might want to review time on a non-digital clock and use the fact that there are 30 degrees between each hour on the clock face. Also, the circumference of this circle of radius 1 is Diameter x π = 2 x π = 6.28…. Therefore, between the small hand and the big hand for 7:55 would be 30 + 30 + 30 + 30 = 120 degrees, and hence the distance on the circumference would be 120/360 of the 6.28 total circumference, i.e., 2.0933.)

Wednesday

1. **Estimation:** 51.2 is 32% of what number? Estimate this number.
   (You might want to tell your students to think of 32% as approximately $\frac{1}{3}$ and to round 51.2 to 50. Therefore you have the question 50 is $\frac{1}{3}$ of what number, and the answer is 150.)

2. **Clock Geometry:** The time is 1:15. What is the angle, and what is the distance?
   (You might want to review time on a non-digital clock and use the fact that there are 30 degrees between each hour on the clock face. Also, the circumference of this circle of radius 1 is Diameter x π = 2 x π = 6.28…. Therefore, between the small hand and the big hand for 1:15 would be 30 + 30 = 60 degrees; hence the distance on the circumference would be 60/360 of the 6.28 total circumference, i.e., 1.04$\overline{6}$.)

29

TEACHER PAGE: QUARTER 1: WEEK 8

Thursday

1. **Calculator Math:** 51.2 is 32% of what number? Calculate this number.

   (You might want to have your students set this up as a proportion $\frac{32}{100} = \frac{51.2}{?}$, which is the same

   as $\frac{100}{32} = \frac{?}{51.2}$, and so $? = 51.2 x \frac{100}{32} = 160$.)

2. **Clock Geometry:** The time is 8:15. What is the angle, and what is the distance?
   (You might want to review time on a non-digital clock and use the fact that there are 30 degrees between each hour on the clock face. Also, the circumference of this circle of radius 1 is Diameter x π = 2 x π = 6.28…. Therefore, between the small hand and the big hand for 8:15 would be 30 + 30 + 30 + 30 + 30 = 150 degrees, and hence the distance on the circumference would be 150/360 of the 6.28 total circumference, i.e., 2.616.)

Friday

1. **Mental Math:** 60 cents for 9 ounces is equal to how many cents for 12 ounces?

   (You might want to suggest setting up a proportion here. In this case, it is $\frac{60}{9} = \frac{20}{3} = \frac{?}{12}$, and the unknown is 80 cents.)

2. **Clock Geometry:** The time is 7:20. What is the angle, and what is the distance?
   (You might want to review time on a non-digital clock and use the fact that there are 30 degrees between each hour on the clock face. Also, the circumference of this circle of radius 1 is Diameter x π = 2 x π = 6.28…. Therefore, between the small hand and the big hand for 7:20 would be 30 + 30 + 30 = 90 degrees, and hence the distance on the circumference would be 90/360 of the 6.28 total circumference, i.e., 1.57.)

Challenge Problems

1. **Clock Geometry:** The time is 1:17. What is the angle, and what is the distance?
   (You might want to review time on a non-digital clock and use the fact that there are 30 degrees between each hour or 6 degrees between each minute on the clock face. Also, the circumference of this circle of radius 1 is Diameter x π = 2 x π = 6.28…. Therefore, between the small hand and the big hand for 1:17 would be 30 + 30 + 12 = 72 degrees, and hence the distance on the circumference would be 72/360 of the 6.28 total circumference, i.e., approximately 1.26.)

2. **Clock Geometry:** The time is 7:08. What is the angle, and what is the distance?
   (You might want to review time on a non-digital clock and use the fact that there are 30 degrees between each hour or 6 degrees between each minute on the clock face. Also, the circumference of this circle of radius 1 is Diameter x π = 2 x π = 6.28…. Therefore, between the small hand and the big hand for 7:08 would be 30 + 30 + 30 + 30 + 30 + 12 = 162 degrees, and hence the distance on the circumference would be 162/360 of the 6.28 total circumference, i.e., approximately 2.826.)

STUDENT PAGE: QUARTER 1: WEEK 8

Student

The theme for week 8 is clock geometry. You will be given a time (assume the hour hand is always right on the number), and you will have to figure out the angle from the big hand to the small hand plus the distance along the circumference from the big hand to the small hand. The radius of the clock is 1 unit. All of the #2 problems will be based on this theme.

Monday

Name:_____

1. **Mental Math:** Add the following numbers: 43 + 32 + 37 + 18.

_____

2. **Clock Geometry:** The time is 10:45. What is the angle, and what is the distance?

_____

Tuesday

Name:_____

1. **Calculator Math:** Calculate the following:
$$\frac{[23 \times (17 + 59) - 147] \times (74 - 36)}{(173 - 52)}$$

_____

2. **Clock Geometry:** The time is 7:55. What is the angle, and what is the distance?

_____

Wednesday

Name:_____

1. **Estimation:** 51.2 is 32% of what number? Estimate this number.

_____

2. **Clock Geometry:** The time is 1:15. What is the angle, and what is the distance?

_____

STUDENT PAGE: QUARTER 1: WEEK 8

Student

Thursday

Name:_____

1. **Calculator Math:** 51.2 is 32% of what number? Calculate this number.

_____

2. **Clock Geometry:** The time is 8:15. What is the angle, and what is the distance?

_____

Friday

Name:_____

1. **Mental Math:** 60 cents for 9 ounces is equal to how many cents for 12 ounces?

_____

2. **Clock Geometry:** The time is 7:20. What is the angle, and what is the distance?

_____

Challenge Problems
          Name:

_____

1. **Clock Geometry:** The time is 1:17. What is the angle, and what is the distance?

_____

2. **Clock Geometry:** The time is 7:08. What is the angle, and what is the distance?

_____

TEACHER PAGE: QUARTER 1: WEEK 9

The theme for week 9 is guess that number. All of the #2 problems are based on this theme. For each of these problems, you will be given a set of word clues. You will then have to become a mathematical detective and figure out what each number is.

Monday

1. **Mental Math:** Add the following fractions: $\frac{2}{7} + \frac{5}{16} + \frac{5}{7}$.

   (You might want to mention that this problem is not very difficult if they see the compatible fractions $\frac{2}{7}$ and $\frac{5}{7}$ that add up to 1. Therefore the answer is $1\frac{5}{16}$.)

2. **Guess That Number:** It is a small deficient (the sum of its proper divisors is less than it) whole number. The sum of its proper divisors is seven, and it is also the cube of a small prime number. Guess that number.

   (You might want to review what proper divisors are and what a cube is. With a little work you will find that for 8 the sum of its proper divisors is $1 + 2 + 4 = 7$. The answer is 8.)

Tuesday

1. **Calculator Math:** Calculate the following: $-2.38 \times [-1.79 + 3.42 \times (-7.39 + 4.26)]$.

   (You might want to tell your students about order of operations and show them how to use the +/- key (or the (-) key) and the parentheses on their calculators. You can perform the following keystrokes: (-) 2.38 x ((-) 1.79 + 3.42 x ((-) 7.39 + 4.26)) =. The answer is 29.7371.)

2. **Guess That Number:** This number represents the ratio of the circumference to the diameter of a circle. It has infinitely many decimal places, and mathematicians have a special symbol for this number. Guess that number.

   (You might want to review the geometry of a circle. You take the formula $C = D \times \pi$ and from that you get $C/D = \pi$. The answer is π or 3.14159265....)

Wednesday

1. **Estimation:** Estimate the following sum: $23.47 + 74.2 + 17.69 + 43.295$.

   (You might instruct your students to round each number to the nearest whole number. In that event, you will get $23 + 74 + 18 + 43 = 158$.)

2. **Guess That Number:** This number has infinitely many decimal places. It is also the length of the diagonal of a square of sides 1. Guess that number.

   (You might want to review the Pythagorean relationship of a right triangle. You would get diagonal$^2 = 1^2 + 1^2 = 2$. The answer is $\sqrt{2}$.)

TEACHER PAGE: QUARTER 1: WEEK 9

Thursday

1. **Calculator Math:** Calculate the following sum: 23.47 + 74.2 + 17.69 + 43.295 without using the decimal point key on your calculator.
    (You might want to review how you add decimal numbers and how the decimal point is not very important. In this case, we would want to perform the following keystrokes: 23470 + 74200 + 17690 + 43295 =. Then put the decimal point three places from the right to get 158.655 as the answer.)

2. **Guess That Number:** This number is the first abundant number (the sum of its proper divisors is bigger than the number) bigger than 30. Guess that number.
    (With a little work you will find that the sum of the proper divisors of 36, 1 + 2 + 3 + 4 + 6 + 9 + 12 + 18 = 55, makes it abundant.)

Friday

1. **Mental Math:** Perform the following division: $\frac{176}{32}$.
    (You might mention that this problem can be attacked by "reducing" it by common factors. It becomes $\frac{176}{32} = \frac{88}{16} = \frac{44}{8} = \frac{22}{4} = \frac{11}{2} = 5\frac{1}{2}$.)

2. **Guess That Number:** This number is a fraction. Both the numerator and denominator are prime numbers, and the percent $66\frac{2}{3}\%$ can be represented by this fraction. Guess that number.
    (You might want to review what % is short for. You can set up
    $$\frac{66\frac{2}{3}}{100} = \frac{\frac{200}{3}}{100} = \frac{200}{300}$$. The answer is $\frac{2}{3}$.)

Challenge Problems

1. **Guess That Number:** This number has infinitely many decimal places. It is the length of the side of a cube whose volume is 2. Guess that number.
    (You might want to review the volume formula for a cube ($V = s \times s \times s$), and then what value of s would satisfy $s \times s \times s = 2$. The answer is the cube root of 2.)

2. **Guess That Number:** This number is the average of the biggest prime number in the 50s and the smallest prime number in the 60s. Guess that number.
    (You might want to review the meaning of prime number. Then with a little figuring you get 59 as the biggest prime in the 50s and 61 as the smallest prime in the 60s. Therefore, the number is 60.)

STUDENT PAGE: QUARTER 1: WEEK 9

Student

The theme for week 9 is guess that number. All of the #2 problems are based on this theme. For each of these problems, you will be given a set of word clues. You will then have to become a mathematical detective and figure out what each number is.

Monday

Name: _____

1. **Mental Math:** Add the following fractions: $\frac{2}{7} + \frac{5}{16} + \frac{5}{7}$. _____

2. **Guess That Number:** It is a small deficient (the sum of its proper divisors is less than it) whole number. The sum of its proper divisors is seven, and it is also the cube of a small prime number. Guess that number.

Tuesday

Name: _____

1. **Calculator Math:** Calculate the following:
   -2.38 x [-1.79 + 3.42 x (-7.39 + 4.26)]. _____

2. **Guess That Number:** This number represents the ratio of the circumference to the diameter of a circle. It has infinitely many decimal places, and mathematicians have a special symbol for this number. Guess that number.

Wednesday

Name: _____

1. **Estimation:** Estimate the following sum:
   23.47 + 74.2 + 17.69 + 43.295. _____

2. **Guess That Number:** This number has infinitely many decimal places. It is also the length of the diagonal of a square of sides 1. Guess that number.

STUDENT PAGE: QUARTER 1: WEEK 9   Student

Thursday   Name: _____

1. **Calculator Math:** Calculate the following sum: _____
    23.47 + 74.2 + 17.69 + 43.295
    without using the decimal point key on your calculator.

2. **Guess That Number:** This number is the first      _____
    abundant number (the sum of its proper divisors
    is bigger than the number) bigger than 30. Guess
    that number.

Friday   Name: _____

1. **Mental Math:** Perform the following division: $\dfrac{176}{32}$.   _____

2. **Guess That Number:** This number is a fraction.      _____
    Both the numerator and denominator are prime
    numbers, and the percent $66\frac{2}{3}\%$ can be repre-
    sented by this fraction. Guess that number.

Challenge Problems   Name: _____

1. **Guess That Number:** This number has infinitely      _____
    many decimal places. It is the length of the side of
    a cube whose volume is 2. Guess that number.

2. **Guess That Number:** This number is the average of the big-   _____
    gest prime number in the 50s and the smallest prime number in
    the 60s. Guess that number.

TEACHER PAGE: QUARTER 2: WEEK 1

The theme of week 1 is travel mathematics. All of the #2 problems are based on this theme.

Monday

1. **Mental Math:** 50 cents for 8 ounces is equal to how many cents for 12 ounces?
   (You might want to mention that this is simply a proportion problem. You can simply mentally set up $\frac{12}{8} = \frac{3}{2} = \frac{?}{50}$ and solve this for 75 cents.)

2. **Travel Math:** If it is 150 miles to Busch Stadium in St. Louis, Missouri, and your parents can average 60 miles per hour, then how long will it take to drive to the stadium?
   (You might want to mention Distance = Rate x Time. In this case you have 150 miles = (60 miles per hour) x Time, which means that Time = $\frac{150}{60}$ hours = $2\frac{1}{2}$ hours.)

Tuesday

1. **Calculator Math:** Perform the following calculation: $\frac{581 \times 39}{45 \times 14 \times 31}$
   (You might have the students all start at a common signal and give them a time limit. Also you might want to tell them the appropriate way to multiply and divide with the calculator without using memory or writing down intermediate answers. You can perform the following keystrokes: 581 x 39 = / 45 = / 14 = / 31 =. The answer is 1.160215054.)

2. **Travel Math:** On Interstate 70, it is 172 miles from Kansas City to Salina, Kansas; 179 miles from Salina to Oakley, Kansas; 57 miles from Oakley to Goodland, Kansas; 29 miles from Goodland to Burlington, Colorado; and finally 160 miles from Burlington to Denver, Colorado. How far is it from Kansas City to Denver?
   (You might want to talk about road maps and the idea that a journey is the sum of smaller trips. Here the answer is just 172 + 179 + 57 + 29 + 160 = 597 miles.)

Wednesday

1. **Estimation:** Estimate (to the nearest hundred) the following sum: 4567 + 538 + 2151 + 309 + 1823.
   (You might give students a hint about rounding each number to the nearest hundred and then adding those numbers. You thus get 4600 + 500 + 2200 + 300 + 1800 = 9400.)

2. **Travel Math:** It is 415 miles from Salt Lake City, Utah, to Las Vegas, Nevada, and you make the trip in $7\frac{1}{4}$ hours. What was your average speed?
   (You need to mention Distance = Rate x Time. In this case you have 415 miles = Rate x $7\frac{1}{4}$ hours. Solving for Rate, you get:
   $$\text{Rate} = \frac{415}{7\frac{1}{4}} = 57 \text{ miles per hour.})$$

37

TEACHER PAGE: QUARTER 2: WEEK 1

Thursday

1. **Calculator Math:** Calculate the average of these test scores: 67, 89, 95, 78, 90, 84, 73, 82. (You might mention how to calculate an average. You can perform the following keystrokes: 67 + 89 + 95 + 78 + 90 + 84 + 73 + 82 = / 8 = . The answer here is 82.25.)

2. **Travel Math:** You notice that your parents are driving on cruise-control at 65 miles per hour. They do this for 3 hours 20 minutes before they decide to stop for a break. How far did you travel?
   (You might mention Distance = Rate x Time. In this case, you simply get Distance = 65 miles per hour x $3\frac{1}{3}$ hours = $216\frac{2}{3}$ miles.)

Friday

1. **Mental Math:** Calculate 28 x 0.2 + 22 x 0.2.
   (You might want to mention using the distributive property. This gives you (28 + 22) x 0.2 = 50 x 0.2 = 10.)

2. **Travel Math:** For the first 100 miles, your parents average 60 miles per hour, but for the next 100 miles, they only average 50 miles per hour. What was their average speed for the 200-mile trip?
   (You might mention Distance = Rate x Time, and in this problem, they need to find the Time for the whole trip by finding the Time for each part of the trip, i.e., $1\frac{2}{3}$ hours + 2 hours = $3\frac{2}{3}$ hours for the whole trip. Thus the Rate = 200 miles / $3\frac{2}{3}$ hours = 54.5 miles per hour.)

Challenge Problem

1. **Travel Math:** You are planning a trip for your parents. The trip is 200 miles and the highways are such that your parents can average 60 miles per hour driving the 200 miles. Your parents want to allow 4 hours total for the trip. How long can you stop for lunch so that the trip is no longer than 4 hours?
   (You will want to use Distance = Rate x Time and figure that the driving time of the trip is Distance divided by Rate or $\frac{200}{60} = 3\frac{1}{3} = 3$ hours 20 minutes. Therefore, you will have 40 minutes left over to stop for lunch.)

STUDENT PAGE: QUARTER 2: WEEK 1   Student

The theme of week 1 is travel mathematics. All of the #2 problems are based on this theme.

- -

Monday   Name:_____

1. **Mental Math:** 50 cents for 8 ounces is equal to how many cents _____
   for 12 ounces?

2. **Travel Math:** If it is 150 miles to Busch Stadium in St. Louis,
   Missouri, and your parents can av-
   erage 60 miles per hour, then how   _____
   long will it take to drive to the stadium?

- -

Tuesday   Name:_____

1. **Calculator Math:** Perform the following calculation: _____

   $$\frac{581 \times 39}{45 \times 14 \times 31}.$$

2. **Travel Math:** On Interstate 70, it is 172 miles from Kansas City
   to Salina, Kansas; 179 miles from Salina to Oakley, Kansas; 57   _____
   miles from Oakley to Goodland, Kansas; 29 miles from Goodland
   to Burlington, Colorado; and finally
   160 miles from Burlington to Denver, Colorado. How far is it from
   Kansas City to Denver?

- -

Wednesday   Name:_____

1. **Estimation:** Estimate (to the nearest hundred) the following sum: _____
   4567 + 538 + 2151 + 309 + 1823.

2. **Travel Math:** It is 415 miles from Salt Lake City, Utah, to Las   _____
   Vegas, Nevada, and you make
   the trip in $7\frac{1}{4}$ hours. What was your
   average speed?

39

STUDENT PAGE: QUARTER 2: WEEK 1

Student

Thursday

Name: _____

1. **Calculator Math:** Calculate the average of these test scores: _____
   67, 89, 95, 78, 90, 84, 73, 82.

2. **Travel Math:** You notice that your parents are driving on cruise- _____
   control at 65 miles per hour. They
   do this for 3 hours 20 minutes
   before they decide to stop for a
   break. How far did you travel?

Friday

Name: _____

1. **Mental Math:** Calculate 28 x 0.2 + 22 x 0.2. _____

2. **Travel Math:** For the first 100 miles, your parents average 60 _____
   miles per hour, but for the next 100 miles, they only average 50
   miles per hour. What was their
   average speed for the 200-mile
   trip?

Challenge Problem

Name: _____

1. **Travel Math:** You are planning a trip for your parents. The trip is _____
   200 miles and the highways are such that your parents can average 60 miles per hour driving the 200 miles. Your parents want
   to allow 4 hours total for the trip. How long can you stop for
   lunch so that the trip is no longer than 4 hours?

TEACHER PAGE: QUARTER 2: WEEK 2

The theme for week 2 is mathematics involved in basketball. All the #2 problems are based on this theme.

Monday

1. **Mental Math:** Multiply the following numbers: 52 x 20.
   (You might mention that the distributive property can be used to make some multiplications easier. For example, the above becomes (50 + 2) x 20 = 50 x 20 + 2 x 20 = 1000 + 40 = 1040.)

2. **Basketball Math:** A player makes 73% of his free throws. At that rate, how many free throws would the player make if he attempts 8 free throws in a game?
   (You might want to review what percentage means in this example. Also, you might explain why you need to round your answer to the nearest whole number. In this problem you simply need to multiply 0.73 x 8 = 5.84. However, a fraction of a free throw doesn't make any sense so round up. The answer is 6 free throws.)

Tuesday

1. **Calculator Math:** Perform the following calculation: [(-23 + 72) x (-45)] / (-34 + 63).
   (You might want to talk about order of operations, parentheses, and the use of the +/- key (or the (-) key) on the calculator. You should emphasize that you should do this problem without using memory or writing down intermediate answers. You can perform the following keystrokes: (((-) 23 + 72) x (-) 45) / ((-) 34 + 63) =. The answer is -76.03448276.)

2. **Basketball Math:** A player played three-fourths of the time in the first half of the game. If the coach wants the player to play the same in the second half, how many of the 16 minutes will she play?
   (You might mention that this is just a simple proportion problem. You can set up the proportion $\frac{3}{4} = \frac{?}{16}$. It easily solves for 12 minutes.)

Wednesday

1. **Estimation:** Estimate the following calculation: 0.00000000626 / 0.00000000000000179.
   (You might want to talk about scientific notation and rounding before doing this problem. You get $6 \times 10^{-9} / 2 \times 10^{-15} = 3 \times 10^{6}$.)

2. **Basketball Math:** A player scored 25 points during the game. He made no free throws. How many two-point and three-point baskets could the player have made to score the 25 points?
   (You might want to mention that there is not just one correct answer. An answer must have an odd number of three-point baskets so that the remainder of points in the 25 is an even number. The interesting thing about this problem is to describe all the possible answers. They are 1-3pt. and 11-2pt.; 3-3pt. and 8-2pt.; 5-3pt. and 5-2pt.; and 7-3pt. and 2-2pt.)

41

TEACHER PAGE: QUARTER 2: WEEK 2

Thursday

1. **Calculator Math:** Perform the following division: 0.00000000626 / 0.00000000000000179. (You might want to show your students how to use scientific notation on their calculators. You can perform the following keystrokes: 6.26 EE (-) 9 / 1.79 EE (-) 15 =. The answer is 3.4972067 EE 6 or x 10^6.)

2. **Basketball Math:** During the season, a player made 23 of her 72 three-point shots. What was her three-point shooting percentage?

   (You might want to review the meaning of percentage. You need to calculate $\frac{23}{72}$ x 100. The answer here is 31.9%.)

Friday

1. **Mental Math:** Perform the following subtraction: 27.57 - 10.96.
   (You might want to tell your students that if you add the same number to both terms in the subtraction problem, you get the same answer. If you add 0.04 to each number, the problem becomes 27.61 - 11 = 16.61.)

2. **Basketball Math:** A player can run 100 meters in 14 seconds. At that rate, how fast could the player run up the basketball court (approximately 30 meters)?
   (You might want to mention that this is a simple proportion problem. You can set up the proportion $\frac{14}{100} = \frac{?}{30}$, which gives you ? $= \frac{14}{100}$ x 30 = 4.2 seconds as the answer.)

Challenge Problems

1. **Sports Math:** The world record for men in the 10,000 meters in track is 26 minutes 22.75 seconds. Use the fact that 10,000 meters is approximately 6.2 miles to figure how long it took this runner to run each mile.
   (You will want to change the time in minutes and seconds into just minutes, so the 26 minutes 22.75 seconds becomes 26.38 minutes. Next, divide this time by the 6.2 miles to get 26.38 / 6.2 = 4.255 minutes, or approximately 4 minutes 15 seconds per mile.)

2. **Sports Math:** The world record for women in the long jump is 7.52 meters. Use the fact that 1 inch is approximately 2.54 centimeters to figure how many feet in 7.52 meters.
   (You will want to convert the meters to centimeters by moving the decimal point two places to the right to become 752 centimeters. Now, divide 752 centimeters by 2.54 to get the number of inches, 752 / 2.54 = approximately 296 inches. Finally, convert this to feet and inches by dividing by 12 to get 24 feet 8 inches.)

STUDENT PAGE: QUARTER 2: WEEK 2

Student

The theme for week 2 is mathematics involved in basketball. All the #2 problems are based on this theme.

Monday

Name: _____

1. **Mental Math:** Multiply the following numbers: 52 x 20.

   _____

2. **Basketball Math:** A player makes 73% of his free throws. At that rate, how many free throws would the player make if he attempts 8 free throws in a game?

   _____

Tuesday

Name: _____

1. **Calculator Math:** Perform the following calculation: [(-23 + 72) x (-45)] / (-34 + 63).

   _____

2. **Basketball Math:** A player played three-fourths of the time in the first half of the game. If the coach wants the player to play the same in the second half, how many of the 16 minutes will she play?

   _____

Wednesday

Name: _____

1. **Estimation:** Estimate the following calculation: 0.00000000626 / 0.00000000000000179.

   _____

2. **Basketball Math:** A player scored 25 points during the game. He made no free throws. How many two-point and three-point baskets could the player have made to score the 25 points?

   _____

STUDENT PAGE: QUARTER 2: WEEK 2

Student

Thursday

Name: _____

1. **Calculator Math:** Perform the following division: 0.00000000626 / 0.00000000000000179.   _____

2. **Basketball Math:** During the season, a player made 23 of her 72 three-point shots. What was her three-point shooting percentage?   _____

Friday

Name: _____

1. **Mental Math:** Perform the following subtraction: 27.57 - 10.96.   _____

2. **Basketball Math:** A player can run 100 meters in 14 seconds. At that rate, how fast could the player run up the basketball court (approximately 30 meters)?   _____

Challenge Problems

Name: _____

1. **Sports Math:** The world record for men in the 10,000 meters in track is 26 minutes 22.75 seconds. Use the fact that 10,000 meters is approximately 6.2 miles to figure how long it took this runner to run each mile.   _____

2. **Sports Math:** The world record for women in the long jump is 7.52 meters. Use the fact that 1 inch is approximately 2.54 centimeters to figure how many feet in 7.52 meters.   _____

TEACHER PAGE: QUARTER 2: WEEK 3

The theme for this week is American Literature. All the #2 problems are based on this theme.

Monday

1. **Mental Math:** Add the following numbers: 47 + 78 + 39.
   (You might want to mention that you can add these mentally by thinking 40 + 70 + 30 = 140 and 7 + 8 + 9 = 24. Therefore the answer is 140 + 24 = 164.)

2. **American Literature:** *The Red Badge of Courage* was a novel of the nineteenth century about the Civil War. It was written by Stephen Crane in the year that has a prime factor of 379. What is the year?
   (You might want to review what prime factor means. Then you can tell that 379 x 2 = 758, 379 x 3 = 1137, and 379 x 4 = 1516 are not the correct answers, but 379 x 5 = 1895 is the correct answer.)

Tuesday

1. **Calculator Math:** Perform the following calculation: [(4.62 + 7.09) x 2.96] / (19.6 - 7.88).
   (You might want to review how to use the parentheses on the calculator so that no memory is used or intermediate answers written down. You can perform the following keystrokes: ((4.62 + 7.09) x 2.96) / (19.6 - 7.88) =. The answer is 2.9574744027.)

2. **American Literature:** In the nineteenth century, before the Civil War, Harriet Beecher Stowe wrote the novel *Uncle Tom's Cabin*. It was about slavery in America. It was first published in the year that is a mere three more than the square of forty-three. What is the year?
   (You might want to talk about translating words into numbers and number operations. The answer is $3 + 43^2 = 1852$.)

Wednesday

1. **Estimation:** At the store you buy two pounds of apples at $0.79 per pound and three pounds of bananas at $0.39 per pound. Estimate the amount of change you might expect if you give the cashier a five-dollar bill.
   (You might want to tell your students to round the prices to the nearest tens, i.e., $0.80 and $0.40 per pound, respectively, so you have about $1.60 of apples and $1.20 of bananas totaling approximately $2.80; hence your change will be approximately $2.20.)

2. **American Literature:** The famous twentieth-century American author Ernest Hemingway wrote the novel *The Old Man and the Sea*. The book was about the struggles of a solitary old man. It was first published in the year where the sum of the units and tens digits is 7 and their product is 10. What is the year?
   (You might want to talk about problem-solving strategies and setting up equations. The equations are $u + t = 7$ and $u \times t = 10$. In this case, the equations are easily solved, and the year is 1952.)

Teacher

Thursday

1. **Calculator Math:** At the store you buy two pounds of apples at $0.79 per pound and three pounds of bananas at $0.39 per pound. Calculate the amount of change you might expect if you give the cashier a five-dollar bill and there is a sales tax of 6%.
   (You might want to tell your student to think of the problem in three parts: calculating a subtotal, calculating the tax and total bill, and calculating the change. You could perform the following keystrokes: 2 x .79 = + 3 x .39 = x 1.06 =. This gives a total of $2.92 (if rounded up). Subtract $2.92 from $5.00 to get $2.08 change.)

2. **American Literature:** The movie *The Last of the Mohicans* was based on the book of the same name written by the early nineteenth-century author James Fenimore Cooper. The novel was first published in the year where the number formed from the tens and units digits is eight more than the number formed from the thousands and hundreds digits. What is the year?
   (You might tell your students what the number formed from the thousands and hundreds digits might look like (since we are talking about the nineteenth century). Then it should be easy to see that the number formed from the tens and units digits is 26, and the year must be 1826.)

Friday

1. **Mental Math:** A student made it to class eight out of ten times in the past two weeks. What percent of classes did the student miss?
   (You might mention that you first have to reason that the student misses class two out of ten days, and thus $\frac{2}{10}$ of the time is absent. Then you can convert $\frac{2}{10}$ to 20%.)

2. **American Literature:** Herman Melville, the nineteenth-century American author, wrote the novel *Moby Dick*. *Moby Dick* is the story of a whale with mystical qualities. The novel was written in the year where the sum of the tens and hundreds digits is 13 and the units and thousands digits are the same. What is the year?
   (You might want to talk about problem-solving strategies. In this problem, it is helpful to look at it as two separate problems: one for finding the units and the thousands digits and the other for finding the tens and hundreds digits. Since the book was written in the nineteenth century, the hundreds digit would be 8 and hence the tens digit must be 5. Likewise, since the thousands digit is 1, then so is the units digit. The answer is 1851.)

Challenge Problem

1. **American Literature:** *The Adventures of Tom Sawyer* is the Samuel Clemens (Mark Twain) novel written about growing up along the Mississippi River in the nineteenth century. This novel was first published in the year whose digits add up to 22 and whose tens digit is one more than its units digit. What is the year?
   (You might want to guess that the book was written in the 1800s (nineteenth century) and so the hundreds and thousands digits are 8 and 1, respectively. That simplifies the problem to a two-digit number the sum of whose digits is 13. Now it should be an easy matter to use the other clue to get the tens digit as 7 and the units digit as 6. Therefore, the year was 1876.)

STUDENT PAGE: QUARTER 2: WEEK 3   **Student**

The theme for this week is American Literature. All the #2 problems are based on this theme.

- -

Monday   Name: _____

1. **Mental Math:** Add the following numbers: 47 + 78 + 39.   _____

2. **American Literature:** *The Red Badge of Courage* was a novel of the nineteenth century about the Civil War. It was written by Stephen Crane in the year that has a prime factor of 379. What is the year?   _____

- -

Tuesday   Name: _____

1. **Calculator Math:** Perform the following calculation: _____
    [(4.62 + 7.09) x 2.96] / (19.6 - 7.88).

2. **American Literature:** In the nineteenth century, before the Civil War, Harriet Beecher Stowe wrote the novel *Uncle Tom's Cabin*. It was about slavery in America. It was first published in the year that is a mere three more than the square of forty-three. What is the year?   _____

- -

Wednesday   Name: _____

1. **Estimation:** At the store you buy two pounds of apples at $0.79 per pound and three pounds of bananas at $0.39 per pound. Estimate the amount of change you might expect if you give the cashier a five-dollar bill.   _____

2. **American Literature:** The famous twentieth-century American author Ernest Hemingway wrote the novel *The Old Man and the Sea*. The book was about the struggles of a solitary old man. It was first published in the year where the sum of the units and tens digits is 7 and their product is 10. What is the year?   _____

STUDENT PAGE: QUARTER 2: WEEK 3   Student

Thursday   Name: _____

1. **Calculator Math:** At the store you buy two pounds of apples at _____
   $0.79 per pound and three pounds of bananas at $0.39 per
   pound. Calculate the amount of change you might expect if you
   give the cashier a five-dollar bill and there is a sales tax of 6%.

2. **American Literature:** The movie *The Last of the Mohicans* was _____
   based on the book of the same name written by the early nineteenth-century author James Fenimore
   Cooper. The novel was first published in
   the year where the number formed from
   the tens and units digits is eight more than
   the number formed from the thousands and
   hundreds digits. What is the year?

Friday   Name: _____

1. **Mental Math:** A student made it to class eight out of ten times in _____
   the past two weeks. What percent of classes did the student
   miss?

2. **American Literature:** Herman Melville, the nineteenth-century _____
   American author, wrote the novel *Moby Dick*. *Moby Dick* is the
   story of a whale with mystical qualities. The
   novel was written in the year where the sum
   of the tens and hundreds digits is 13 and the
   units and thousands digits are the same.
   What is the year?

Challenge Problem   Name: _____

1. **American Literature:** *The Adventures of Tom Sawyer* is the _____
   Samuel Clemens (Mark Twain) novel written about growing up
   along the Mississippi River in the nineteenth
   century. This novel was first published in the
   year whose digits add up to 22 and whose tens
   digit is one more than its units digit. What is
   the year?

TEACHER PAGE: QUARTER 2: WEEK 4

The theme for week 4 is American presidents. All the #2 problems are based on this theme.

Monday

1. **Mental Math:** Find $\frac{7}{12}$ of 240.

   (You might mention to the students that this can be done by thinking of it as finding $\frac{1}{12}$ of 240 and then multiplying by 7, i.e., 7 x 20 = 140.)

2. **American Presidents:** Thomas Jefferson was the third President of the United States. He became president in the year whose sum of digits is 10 and where the units digit is greater than the tens digit.

   (You might want to talk about problem-solving strategies and setting up equations. You would get $th + h + te + u = 10$ and $u > te$. Since the hundreds digit is either 7 or 8, you can guess-and-check to figure out that the year is 1801.)

Tuesday

1. **Calculator Math:** Multiply the following numbers without using the decimal point on your calculator: 57 x 7.83.

   (You might want to review how we multiply numbers by hand and then explain how this might help students determine where the decimal point goes in the above problem. You can perform the following keystrokes: 57 x 783 = to get 44631, and then move the decimal point two places from the right. The answer is 446.31.)

2. **American Presidents:** Woodrow Wilson was president from 1913 to 1921. He was the nth president, where n is the number that is one more than the cube of 3. What number president was Woodrow Wilson?

   (You might want to review translating words into mathematical operations. In this case $n = 3^3 + 1 = 28$).

Wednesday

1. **Estimation:** Estimate 35% of $115.76.

   (You might want to tell your students that there are several ways to make this estimation. You could think of 35% as approximately $\frac{1}{3}$ and $115.76 as approximately $120 and get $40.)

2. **American Presidents:** John Quincy Adams was the sixth President of the United States. He served only one term in office, which began in a year that is divisible by 25. What year did Adams become president?

   (You might want to talk about what it means to be divisible by 25. Then you will know that the year is one of the following, 1775, 1800, 1825, 1850, …. With a little knowledge of history, you can figure out that the correct answer must be 1825.)

49

TEACHER PAGE: QUARTER 2: WEEK 4

Thursday

1. **Calculator Math:** Calculate 35% of $115.76 without using the percent key on your calculator. Try it without even using the decimal point.
   (You might want to review what percent means. If you do it without using the decimal point on the calculator, then you multiply 35 x 11576 to get 405160. However, you know your answer should be about 40, so the answer must be 40.516.)

2. **American Presidents:** Ulysses S. Grant, the Civil War general, was the nth President of the United States. The number n is a two-digit number the sum of whose digits is 9 and the product is 8. What is n?
   (You might want to review problem-solving strategies and setting up equations. The equations would be $t + u = 9$ and $t \times u = 8$. By guess-and-check, you should be able to find that $n = 18$ or 81. But 81 cannot be the correct answer, since there have only been 43 presidents, so Grant must be the 18th president.)

Friday

1. **Mental Math:** If a dozen donuts cost $3.60, then how much will 30 donuts cost?
   (You might want to mention that one way of doing this problem is to figure out how much one donut costs ($\frac{1}{12}$ of $3.60 = $0.30), and then multiply that amount by 30; thus the answer is $9.00.)

2. **American Presidents:** John F. Kennedy was the 35th President of the United States. He served only part of one term before he was assassinated. The year that he went into office is divisible by the prime number 37. It is also divisible by another prime. In what year did Kennedy become president?
   (You might want to review what a prime number is and also what it means to be divisible by a number. Then you know that the year must be one of the following: 37 x 52 = 1924, 37 x 53 = 1961, or 37 x 54 = 1998. However, the only other factor to be prime is 53; thus the year is 1961.)

Challenge Problem

1. **American Presidents:** James Madison of Virginia helped draft the Bill of Rights and served in President Jefferson's cabinet as Secretary of State. Madison himself became President of the United States in the year whose sum of digits is 18 and whose units digit is equal to the sum of the thousands and hundreds digit. What is the year?
   (You can probably guess that Madison was president in the 1800s. So the rest should be easy. With the other clue, the units digit must be 9 and since the sum of all the digits is 18, then the tens digit must be 0; so the answer is 1809.)

STUDENT PAGE: QUARTER 2: WEEK 4

The theme for week 4 is American presidents. All the #2 problems are based on this theme.

Monday   Name: _____

1. **Mental Math:** Find $\frac{7}{12}$ of 240.   _____

2. **American Presidents:** Thomas Jefferson was the third President of the United States. He became president in the year whose sum of digits is 10 and where the units digit is greater than the tens digit.   _____

T. JEFFERSON

Tuesday   Name: _____

1. **Calculator Math:** Multiply the following numbers without using the decimal point on your calculator: 57 x 7.83.   _____

2. **American Presidents:** Woodrow Wilson was president from 1913 to 1921. He was the *n*th president, where *n* is the number that is one more than the cube of 3. What number president was Woodrow Wilson?   _____

W. WILSON

Wednesday   Name: _____

1. **Estimation:** Estimate 35% of $115.76.   _____

2. **American Presidents:** John Quincy Adams was the sixth President of the United States. He served only one term in office, which began in a year that is divisible by 25. What year did Adams become president?   _____

J.Q. ADAMS

51

STUDENT PAGE: QUARTER 2: WEEK 4

Student

Thursday   Name: _____

1. **Calculator Math:** Calculate 35% of $115.76 without using the _____
   percent key on your calculator. Try it without even using the decimal point.

2. **American Presidents:** Ulysses S. Grant, the _____
   Civil War general, was the *n*th President of the
   United States. The number *n* is a two-digit number the sum of whose digits is 9 and the product is 8. What is *n*?

U.S. GRANT

Friday   Name: _____

1. **Mental Math:** If a dozen donuts cost $3.60, then how much will _____
   30 donuts cost?

2. **American Presidents:** John F. Kennedy was the _____
   35th President of the United States. He served
   only part of one term before he was assassinated. The year that he went into office is divisible by the prime number 37. It is also divisible by another prime. In what year did Kennedy
   become president?

J. KENNEDY

Challenge Problem   Name: _____

1. **American Presidents:** James Madison of Virginia _____
   helped draft the Bill of Rights and served in
   President Jefferson's cabinet as Secretary of
   State. Madison himself became President of
   the United States in the year whose sum of
   digits is 18 and whose units digit is equal to
   the sum of the thousands and hundreds digit.
   What is the year?

J. MADISON

TEACHER PAGE: QUARTER 2: WEEK 5

The theme for week 5 is ancient Chinese and Hindu mathematics. All of the #2 problems are based on this theme.

Monday

1. **Mental Math:** Add 23.7 + 15.8 + 31.3.
   (You might want to mention grouping compatible numbers. For example, 23.7 + 31.3 = 55 and so our sum becomes 55 + 15.8 = 70.8.)

2. **Chinese Math:** It is possible that this magic square dates from around 2200 B.C.E. Can you complete it?

4	9	2

   (You might want to mention that a magic square is a square array of distinct integers that have the same sum on any row, column, or main diagonal. The answer here is

4	9	2
3	5	7
8	1	6

   .)

Tuesday

1. **Calculator Math:** Perform the calculation: [23 + (73 - 59) x 82] / (112 + 39).
   (You might want to talk about order of operations and use of parentheses. You can perform the following keystrokes: (23 + (73 - 59) x 82) / (112 + 39) =. The answer is 7.754966887.)

2. **Chinese Math:** This problem dates back to ancient China. When each person in a group (of unknown size) chips in $8 to buy some item, they are $3, in total, over the price of the item. If each person in the same group instead chips in $7 to buy the same item, they are $4, in total, under the price of the item. How many people are in the group, and what is the price of the item?
   (You might want to set up an equation that equates the total price of the item using n as the number of people in the group. You would get 8 x n - 3 = Total Price = 7 x n + 4. This can be solved to find $n = 7$. So the price is $53, and there are 7 people in the group.)

Wednesday

1. **Estimation:** Estimate $\dfrac{(4.82 \times 10^7) \times (8.39 \times 10^6)}{6.72 \times 10^8}$
   (You might want to talk about scientific notation. A good approximation would be
   $\dfrac{5 \times 8 \times 10^{13}}{7 \times 10^8} = 5.7 \times 10^5$.)

TEACHER PAGE: QUARTER 2: WEEK 5

Wednesday (continued)

2. **Hindu Math:** This Hindu problem dates back to around 850 C.E. You are asked to find out how many mangoes were present at the beginning if the first prince took $\frac{1}{4}$, the second prince took $\frac{1}{3}$ of the remaining, and finally the third prince took $\frac{1}{2}$ of those left, leaving 3 mangoes.
(You might want to mention that it is good to start at the end and work back to the beginning in some problems. Since the third prince took $\frac{1}{2}$ and that left 3 then he must have taken 3 himself. Thus there were 6 before the third prince picked. Now the second prince took $\frac{1}{3}$ and thus left $\frac{2}{3}$, which is the 6. So there must have been 9 before the second prince picked. Likewise, you can deduce that there must have been 12 mangoes to begin with. The answer is 12 mangoes.)

Thursday

1. **Calculator Math:** Calculate the following: $\dfrac{(4.82 \times 10^7) \times (8.39 \times 10^6)}{6.72 \times 10^8}$

(You might want to mention how to express numbers in scientific notation on a calculator. You can perform the following keystrokes: 4.82 EE 7 x 8.39 EE 6 = / 6.72 EE 8 =. The answer is 6.017827381×10^5.)

2. **Hindu Math:** Hindu mathematicians over 1500 years ago solved problems like this one. In how many ways can the sum of 5 dollars be paid in dimes and quarters?
(You might want to mention that there is not one way here, but many ways. You can give 0 quarters and 50 dimes, 2 quarters and 45 dimes, 4 quarters and 40 dimes, …, finally 20 quarters and 0 dimes.)

Friday

1. **Mental Math:** Multiply $2\frac{1}{7} \times 28$.
(You might want to mention how the distributive property might be useful in this problem. Here $(2 + \frac{1}{7}) \times 28 = 2 \times 28 + \frac{1}{7} \times 28 = 56 + 4 = 60$.)

2. **Hindu Math:** Early Hindu mathematicians probably multiplied two numbers by a process called lattice multiplication. For example, 24 x 63 can be multiplied by making the lattice as below and then adding down the diagonals from right to left, carrying where necessary.

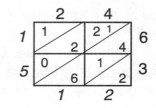

Use lattice multiplication to multiply 374 x 49.
(You might want to go over the example above. The correct lattice looks like:

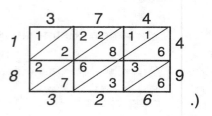

.)

54

STUDENT PAGE: QUARTER 2: WEEK 5

The theme for week 5 is ancient Chinese and Hindu mathematics. All of the #2 problems are based on this theme.

Monday

Name: _____

1. **Mental Math:** Add 23.7 + 15.8 + 31.3.   _____

2. **Chinese Math:** It is possible that this magic square dates from around 2200 B.C.E. Can you complete it?

4	9	2

Tuesday

Name: _____

1. **Calculator Math:** Perform the calculation: _____
   [23 + (73 - 59) x 82] / (112 + 39).

2. **Chinese Math:** This problem dates back to ancient China. When   _____
   each person in a group (of unknown size) chips in $8 to buy
   some item, they are $3, in total, over the price of the item. If
   each person in the same group instead chips in $7 to buy the
   same item, they are $4, in total, under the price of the item. How
   many people are in the group, and what is the price of the item?

Wednesday

Name: _____

1. **Estimation:** Estimate $\dfrac{(4.82 \times 10^7) \times (8.39 \times 10^6)}{6.72 \times 10^8}$ _____

2. **Hindu Math:** This Hindu problem dates back to around 850 C.E.   _____
   You are asked to find out how many mangoes were present at
   the beginning if the first prince took $\frac{1}{4}$, the second prince took $\frac{1}{3}$ of
   the remaining, and finally the third prince took $\frac{1}{2}$ of those left,
   leaving 3 mangoes.

STUDENT PAGE: QUARTER 2: WEEK 5

Student

Thursday   Name: _____

1. **Calculator Math:** Calculate the following: _____

    $$\frac{(4.82 \times 10^7) \times (8.39 \times 10^6)}{6.72 \times 10^8}$$

2. **Hindu Math:** Hindu mathematicians over 1500 years ago solved   _____
   problems like this one. In how many ways can the sum of 5
   dollars be paid in dimes and quarters?

Friday   Name: _____

1. **Mental Math:** Multiply $2\frac{1}{7} \times 28$.   _____

2. **Hindu Math:** Early Hindu mathematicians probably multiplied   _____
   two numbers by a process called lattice multiplication. For example, 24 x 63 can be multiplied by making the lattice as below
   and then adding down the diagonals from right to left, carrying
   where necessary.

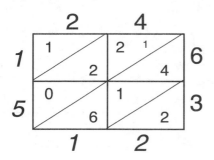

Use lattice multiplication to multiply 374 x 49.

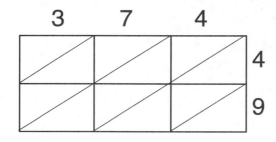

56

TEACHER PAGE: QUARTER 2: WEEK 6

The theme for week 6 is famous national parks and their zip codes. We will think of a zip code as being composed of a two-digit number, followed by a one-digit number, followed by another two-digit number, e.g., 63435 would be 63, 4, and 35. All of the #2 problems are based on this theme.

Monday

1. **Mental Math:** Perform the following division: 6840 / 100.
   (You might want to mention that dividing by a power of ten is accomplished by simply moving the decimal place of the dividend by an appropriate number of places. In our case, move the decimal point just two places to the left, making the answer 68.40.)

2. **National Parks:** The Statue of Liberty is probably the most famous symbol of the United States. It sits on Liberty Island in New York, and it was a gift from France. It was dedicated in 1886. The sum of the three numbers in its zip code is 14, and the first number is 6 more than the third. The middle number is nothing. What is the zip code?
   (You might want to mention problem-solving strategies and setting up equations. The equations might be $f + s + t = 14$ and $f = t + 6$. But since the second, s, is 0, then it is easy to guess-and-check that $f = 10$ and $t = 4$. The answer is 10004.)

Tuesday

1. **Calculator Math:** Multiply: $27\frac{1}{3} \times 8\frac{3}{4}$.

   (You might want to discuss the difference between solving it by converting to decimals, 27.33 x 8.75, and converting to simple fractions, $\frac{82}{3} \times \frac{35}{4}$. In the first case, you are having to use a rounded number (27.33) and so your answer is at best an approximation. In the second case, you get $239.1\overline{6}$, the exact answer.)

2. **National Parks:** The first national park in the United States was Yellowstone Park, which was created by Congress in 1872. It consists of 3,400 square miles of some of the most remarkable land in America, including natural wonders like Old Faithful. The first number in the zip code is one more than nine squared. The average of the first and third number is 86, and the second number is a factor of every whole number. What is the zip code?
   (You might want to review what average means and also what it means to be a factor. The first number must be 1 + 9 x 9 = 82, and since the average of the first and last is 86, then the last number must be 90. The number that is a factor of every whole number is 1. The answer is 82190.)

Wednesday

1. **Estimation:** If an item costs $4.57 each, then estimate the cost of two dozen of the items.
   (You might suggest rounding the price to the nearest dollar and then multiplying. If you did that you would get $5 x 24 = $120.)

2. **National Parks:** Mount Rushmore in South Dakota is another famous symbol of America. It is the work of Gutzon Borglum who sculpted the four faces (George Washington, Thomas Jefferson, Theodore Roosevelt, and Abraham Lincoln) on Mount Rushmore between 1927 and 1941. The middle number in its zip code is the fourth prime number. The first and third numbers have

TEACHER PAGE: QUARTER 2: WEEK 6

Wednesday (continued)

factors of 19 and 17, respectively, and their average is 54. What is the zip code of Mount Rushmore?

(You might want to review average of two numbers and what it means to be a factor. Also, remember that you start counting primes from 2. The first and third numbers might be 38 and 34, 57 and 51, or 76 and 68. However, 57 and 51 have the desired average of 54. Starting from 2, the primes are 2, 3, 5, 7, 11, etc. Therefore, the fourth prime is 7. The answer is 57751.)

Thursday

1. **Calculator Math:** If an item costs $4.57 each, then how much would two dozen of the items cost?

    (You might want to review what is meant by dozen. Otherwise, it is simply $4.57 x 24 = $109.68.)

2. **National Parks:** Gettysburg Battlefield is another famous national park. It was the site of the most infamous battle of the Civil War. On July 1, 1863, the three-day battle began. At the end of the battle, 51,000 soldiers had been killed, wounded, or captured. Later that same year Gettysburg was the site of Abraham Lincoln's famous "Gettysburg Address." The sum of the three numbers of the zip code is 45. The third number is the square of five, and the second number is the first odd prime. What is the zip code of Gettysburg?

    (You might want to talk about problem-solving strategies and review what a square number is. Since the third number must be 25 and the second number 3, the first number must be 17. The answer is 17325.)

Friday

1. **Mental Math:** Perform the following multiplication: $24 \times 3\frac{1}{3}$.

    (You might want to review the distributive rule here. In this problem you have $24 \times (3 + \frac{1}{3}) = 24 \times 3 + 24 \times \frac{1}{3} = 72 + 8 = 80$.)

2. **National Parks:** The Grand Canyon is one of the most striking of all natural landscapes in America. The first number in the zip code is five more than nine squared and the third number is two less than five squared. The sum of the three numbers is 109. What is the zip code?

    (You might want to review translating words into numbers and number operations. Since the first number is $5 + (9 \times 9) = 86$ and the third number is $5 \times 5 - 2 = 23$, then the second number must be 0. The answer here is 86023.)

Challenge Problem

1. **Calculator Math:** When you divide 4 into 9, you get 2 with a remainder of 1. However, if you do this on your calculator, you get (unless you use the integer divide key) 2.25. Use your calculator to find the quotient and remainder when you divide 2487 by 43. Do not use the integer divide key if you have one (except to check your answer).

    (You might want to show the students that to get the whole number remainder, you take the decimal fraction part of the answer and multiply by the divisor. In this problem, when you divide 2487 / 43, you get 57.8371. So 57 is your quotient and the .8372 is the fraction of 43 that represents your remainder. So 43 x 0.8372 = 36 and 36 is the remainder.)

STUDENT PAGE: QUARTER 2: WEEK 6   Student

The theme for week 6 is famous national parks and their zip codes. We will think of a zip code as being composed of a two-digit number, followed by a one-digit number, followed by another two-digit number, e.g., 63435 would be 63, 4, and 35. All of the #2 problems are based on this theme.

Monday   Name: _____

1. **Mental Math:** Perform the following division: 6840 / 100.   _____

2. **National Parks:** The Statue of Liberty is probably the most   _____
   famous symbol of the United States. It sits on Liberty Island
   in New York, and it was a gift from France. It was dedicated
   in 1886. The sum of the three numbers in its zip code is 14,
   and the first number is 6 more than the third. The middle number is nothing. What is the zip code?

Tuesday   Name: _____

1. **Calculator Math:** Multiply: $27\frac{1}{3} \times 8\frac{3}{4}$.   _____

2. **National Parks:** The first national park in the United States was   _____
   Yellowstone Park, which was created by Congress in 1872. It
   consists of 3,400 square miles of some of the most remarkable
   land in America, including natural wonders like Old Faithful. The
   first number in the zip code is one more than nine squared. The
   average of the first and third number is 86, and the second number is a factor of every whole number. What is the zip code?

Wednesday   Name: _____

1. **Estimation:** If an item costs $4.57 each, then estimate the cost   _____
   of two dozen of the items.

2. **National Parks:** Mount Rushmore in South Dakota is another   _____
   famous symbol of America. It is the work of Gutzon Borglum
   who sculpted the four faces (George Washington, Thomas
   Jefferson, Theodore Roosevelt, and Abraham Lincoln) on Mount
   Rushmore between 1927 and 1941. The middle number in its
   zip code is the fourth prime number. The first and third numbers
   have factors of 19 and 17, respectively, and their average is 54.
   What is the zip code of Mount Rushmore?

STUDENT PAGE: QUARTER 2: WEEK 6

Student

Thursday

Name: _____

1. **Calculator Math:** If an item costs $4.57 each, then how much would two dozen of the items cost?

    _____

2. **National Parks:** Gettysburg Battlefield is another famous national park. It was the site of the most infamous battle of the Civil War. On July 1, 1863, the three-day battle began. At the end of the battle, 51,000 soldiers had been killed, wounded, or captured. Later that same year Gettysburg was the site of Abraham Lincoln's famous "Gettysburg Address." The sum of the three numbers of the zip code is 45. The third number is the square of five, and the second number is the first odd prime. What is the zip code of Gettysburg?

    _____

Friday

Name: _____

1. **Mental Math:** Perform the following multiplication: $24 \times 3\frac{1}{3}$.

    _____

2. **National Parks:** The Grand Canyon is one of the most striking of all natural landscapes in America. The first number in the zip code is five more than nine squared and the third number is two less than five squared. The sum of the three numbers is 109. What is the zip code?

    _____

Challenge Problems

Name: _____

1. **Calculator Math:** When you divide 4 into 9, you get 2 with a remainder of 1. However, if you do this on your calculator, you get (unless you use the integer divide key) 2.25. Use your calculator to find the quotient and remainder when you divide 2487 by 43. Do not use the integer divide key if you have one (except to check your answer).

    _____

TEACHER PAGE: QUARTER 2: WEEK 7

The theme for week 7 will be making change. You will be given the amount of a purchase and the amount of money given to pay for the purchase. You are to figure the change to give back by using the least amount of bills and coins in your cash register. The register contains the usual coins (pennies, nickels, dimes, and quarters) and the usual bills (ones, fives, tens, and twenties). All of the #2 problems are based on this theme.

Monday

1. **Mental Math:** Perform the following multiplication: 28 x 12.
   (You might want to mention that the distributive property makes this problem possible for a mental calculation. It becomes 28 x (10 + 2) = 28 x 10 + 28 x 2 = 280 + 56 = 336.)

2. **Make Change:** The purchase is $0.18, and the amount tendered is $1.00.
   (You might suggest that your students try making change mentally by adding up to the amount tendered. Then they could check it with pencil-and-paper methods. The answer is 2-P, 1-N, and 3-Q.)

Tuesday

1. **Calculator Math:** Add the following list of numbers without using the decimal point key on your calculator: $22,849.45 + $35,006.28 + $8,937.95 + $101,317.82.
   (You might want to explain why it isn't necessary to use the decimal place key in the calculation. You can perform the following keystrokes: 2284945 + 3500628 + 893795 + 10131782 = to get 16811150, and then just put the decimal point two places from the right. The answer is $168,111.50.)

2. **Make Change:** The purchase is $4.67, and the amount tendered is $10.00.
   (You might suggest that your students try making change mentally by adding up to the amount tendered. Then they could check it with pencil-and-paper methods. The answer is 3-P, 1-N, 1-Q, and 1-$5.)

Wednesday

1. **Estimation:** Estimate the following calculation involving scientific notation:
   $$\frac{(4.78 \times 10^6) \times (8.19 \times 10^3)}{6.21 \times 10^7}$$
   (You might want to mention that you can round the numbers and operate in any order. It would probably make sense to do the powers of ten last. You would get
   $$\frac{5 \times 8 \times 10^6 \times 10^3}{6 \times 10^7} = \frac{40 \times 10^9}{6 \times 10^7} = 6.7 \times 10^2.)$$

2. **Make Change:** The purchase is $4.38, and the amount tendered is $10.03.
   (You might suggest that your students try making change mentally by adding up to the amount tendered. Then they could check it with pencil-and-paper methods. The answer is 1-N, 1-D, 2-Q, and 1-$5.)

TEACHER PAGE: QUARTER 2: WEEK 7   Teacher

Thursday

1. **Calculator Math:** Perform the following calculation involving scientific notation:

$$\frac{(4.78 \times 10^6) \times (8.19 \times 10^3)}{6.21 \times 10^7}$$

   (You might want to talk about the order of doing the calculation and how to use scientific notation on a calculator. You can perform the following keystrokes: 4.78 EE 6 x 8.19 EE 3 = / 6.21 EE 7 =. The answer is 630.406.)

2. **Make Change:** The purchase is $11.31, and the amount tendered is $21.50.
   (You might suggest that your students try making change mentally by adding up to the amount tendered. Then they could check it with pencil-and-paper methods. The answer is 4-P, 1-N, 1-D, and 1-$10.)

Friday

1. **Mental Math:** Calculate $\frac{3}{7}$ of 770.

   (You might want to mention that it is possible to think of this as $3 \times \frac{1}{7} \times 770$, and hence this becomes 3 x 110 = 330.)

2. **Make Change:** The purchase is $36.42, and the amount tendered is $100.00.
   (You might suggest that your students try making change mentally by adding up to the amount tendered. Then they could check it with pencil-and-paper methods. The answer is 3-P, 1-N, 2-Q, 3-$1, and 3-$20.)

Challenge Problems

1. **Make Change:** The purchase is $121.43, and the amount tendered is $150.00.
   (You might suggest that your students try making change mentally by adding up to the amount tendered. Then they could check it with pencil-and-paper methods. The answer is 2-P, 1-N, 2-Q, 3-$1, 1-$5, and 1-$20.)

2. **Make Change:** The purchase is $111.61, and the amount tendered is $122.11.
   (You might suggest that your students try making change mentally by adding up to the amount tendered. Then they could check it with pencil-and-paper methods. The answer is 2-Q and 1-$10.)

3. **Make Change:** The purchase is $83.79, and the amount tendered is $104.04.
   (You might suggest that your students try making change mentally by adding up to the amount tendered. Then they could check it with pencil-and-paper methods. The answer is 1-Q and 1-$20.)

STUDENT PAGE: QUARTER 2: WEEK 7

Student

The theme for week 7 will be making change. You will be given the amount of a purchase and the amount of money given to pay for the purchase. You are to figure the change to give back by using the least amount of bills and coins in your cash register. The register contains the usual coins (pennies, nickels, dimes, and quarters) and the usual bills (ones, fives, tens, and twenties). All of the #2 problems are based on this theme.

Monday

Name: _____

1. **Mental Math:** Perform the following multiplication: 28 x 12.   _____

2. **Make Change:** The purchase is $0.18, and the amount tendered is $1.00.   _____

Tuesday

Name: _____

1. **Calculator Math:** Add the following list of numbers without using the decimal point key on your calculator: _____
   $22,849.45 + $35,006.28 + $8,937.95 + $101,317.82.

2. **Make Change:** The purchase is $4.67, and the amount tendered is $10.00.   _____

Wednesday

Name: _____

1. **Estimation:** Estimate the following calculation involving scientific notation: _____
   $$\frac{(4.78 \times 10^6) \times (8.19 \times 10^3)}{6.21 \times 10^7}.$$

2. **Make Change:** The purchase is $4.38, and the amount tendered is $10.03.   _____

STUDENT PAGE: QUARTER 2: WEEK 7 — Student

Thursday   Name: _____

1. **Calculator Math:** Perform the following calculation involving _____
   scientific notation: $\dfrac{(4.78 \times 10^6) \times (8.19 \times 10^3)}{6.21 \times 10^7}$.

2. **Make Change:** The purchase is $11.31, and the   _____
   amount tendered is $21.50.

Friday   Name: _____

1. **Mental Math:** Calculate $\dfrac{3}{7}$ of 770.   _____

2. **Make Change:** The purchase is $36.42, and the   _____
   amount tendered is $100.00.

Challenge Problems   Name: _____

1. **Make Change:** The purchase is $121.43, and the amount ten-   _____
   dered is $150.00.

2. **Make Change:** The purchase is $111.61, and the amount ten-   _____
   dered is $122.11.

3. **Make Change:** The purchase is $83.79, and the   _____
   amount tendered is $104.04.

64

TEACHER PAGE: QUARTER 2: WEEK 8

The theme for week 8 is triangle, rectangle, and circle geometry. You will be given certain important dimensions of a shape that is composed of triangles, rectangles, and circles and asked to find the area of the shape. All of the #2 problems are based on this theme.

Monday

1. **Mental Math:** Perform the following calculation: 28 - 34 + 17 - 26 + 73.
   (You might want to mention that if there are compatible numbers, then it might be a good idea to combine them first. Thus in our problem we would have 28 - 60 + 90 = 28 + 30 = 58.)

2. **T-R-C Geometry:** Find the area of the whole region shown.
   (You might want to review the area of triangles, rectangles, and circles and have your students look at the whole as the sum of its components. The area of the triangle is $\frac{1}{2}$ of 3 x 4, which is 6, and the area of the square is 4 x 4 = 16. The answer is 22.)

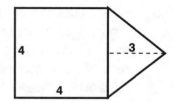

Tuesday

1. **Calculator Math:** Calculate: $\dfrac{1}{\sqrt{13}+\sqrt{8}}$.
   (You might want to discuss using the square root key and the reciprocal key. You can perform the following keystrokes: √ 13 + √ 8 = x⁻¹ =. The answer is 0.155425.)

2. **T-R-C Geometry:** Find the area of the whole region shown.
   (You might want to review the area of triangles, rectangles, and circles and have your students look at the whole as the sum of its components. The area of the rectangle is 4 x 7 = 28, the area of the semi-circle is $\frac{1}{2}$ of R^2 x π or just $\frac{1}{2}$ of 4π. The answer is 34.28.)

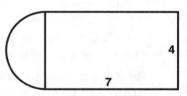

Wednesday

1. **Estimation:** Estimate 639.23 / 20.3.
   (You might want to talk about rounding to appropriate numbers. In this case, 640 / 20 = 32.)

2. **T-R-C Geometry:** Find the area of the whole region shown.
   (You might want to review the area of triangles, rectangles, and circles and have your students look at the whole as the sum of its components. The area of the rectangle is 3 x 5 = 15, and the area of each triangle is $\frac{1}{2}$ of 2 x 3, which is 3. The answer is 21).

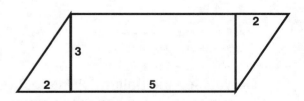

       65

TEACHER PAGE: QUARTER 2: WEEK 8

Thursday

1. **Calculator Math:** Calculate 639.23 / 20.3 without using the decimal point key on your calculator.
   (You might want to review the division algorithm, and how the actual division doesn't depend on the decimal place. The student just needs to keep track of where it should be for the answer. You can perform the following keystrokes: 63923 / 203 = and get 314.892. Then, since there is one decimal place in the divisor, you must move the decimal point one more place to the left in the calculator answer. Our answer here is 31.4892.)

2. **T-R-C Geometry:** Find the area of the whole region shown.
   (You might want to review the area of triangles, rectangles, and circles and have your students look at the whole as the sum of its components. The area of the rectangle is 4 x 8 = 32, and the area of the triangle is $\frac{1}{2}$ of 2 x 8 or 8. The answer is 40).

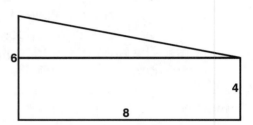

Friday

1. **Mental Math:** Multiply: 7 x 11.8.
   (You might mention that the distributive law would help in this situation. It makes the problem 7 x (11 + 0.8) = 7 x 11 + 7 x 0.8 = 77 + 5.6 = 82.6.)

2. **T-R-C Geometry:** Find the area of the whole region shown.
   (You might want to review the area of triangles, rectangles, and circles and have your students look at the whole as the sum of its components. The area of the square in the middle is 3 x 3 = 9, and the area of each of the four triangles is $\frac{1}{2}$ of 2 x 5, which is 5. The answer is 29).

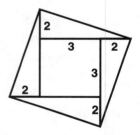

Challenge Problem

1. **Calculator Math:** Use your calculator to compute each of the following:
   1 = _____
   1 - 4 = _____
   1 - 4 + 9 = _____
   1 - 4 + 9 - 16 = _____
   1 - 4 + 9 - 16 + 25 = _____.
   Do you see the pattern? Write out the next three terms in the pattern.
   (You might want to have the students look at the difference between the whole number parts of the successive terms in the sequence of answers: 1, -3, 6, -10, 15. That is 3 - 1 = 2, 6 - 3 = 3, 10 - 6 = 4, and 15 - 10 = 5. With this they should be able to figure out that the next three terms are -21, 28, and -36.)

STUDENT PAGE: QUARTER 2: WEEK 8

Student

The theme for week 8 is triangle, rectangle, and circle geometry. You will be given certain important dimensions of a shape that is composed of triangles, rectangles, and circles and asked to find the area of the shape. All of the #2 problems are based on this theme.

Monday   Name:_____

1. **Mental Math:** Perform the following calculation: _____
   28 - 34 + 17 - 26 + 73.

2. **T-R-C Geometry:** Find the area of the whole region shown.   _____

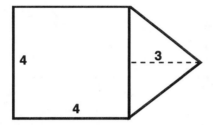

Tuesday   Name:_____

1. **Calculator Math:** Calculate: $\dfrac{1}{\sqrt{13}+\sqrt{8}}$.   _____

2. **T-R-C Geometry:** Find the area of the whole region shown.   _____

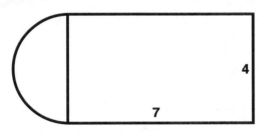

Wednesday   Name:_____

1. **Estimation:** Estimate 639.23 / 20.3.   _____

2. **T-R-C Geometry:** Find the area of the whole region shown.   _____

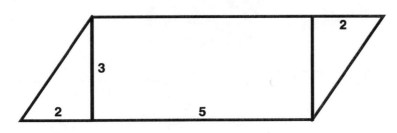

STUDENT PAGE: QUARTER 2: WEEK 8

Student

Thursday Name: _____

1. **Calculator Math:** Calculate 639.23 / 20.3 without using the decimal point key on your calculator.   _____

2. **T-R-C Geometry:** Find the area of the whole region shown.   _____

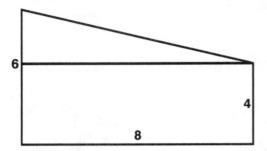

Friday Name: _____

1. **Mental Math:** Multiply: 7 x 11.8.   _____

2. **T-R-C Geometry:** Find the area of the whole region shown.   _____

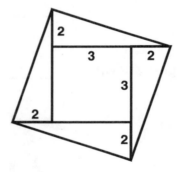

Challenge Problem Name: _____

1. **Calculator Math:** Use your calculator to compute each of the following:

   1 = _____

   1 - 4 = _____

   1 - 4 + 9 = _____

   1 - 4 + 9 - 16 = _____

   1 - 4 + 9 - 16 + 25 = _____.

   Do you see the pattern? Write out the next three terms in the pattern.   _____

TEACHER PAGE: QUARTER 2: WEEK 9

The theme for week 9 is guess that shape. All of the #2 problems are based on this theme. For each of these problems, you will be given a set of clues. You will then have to become a mathematical detective to figure out the most descriptive name for the plane polygonal shape.

Monday

1. **Mental Math:** Add the following decimals: 5.8 + 3.5 + 6.7.
   (You might want to instruct your students to first add the whole numbers 5 + 3 + 6 = 14 and then add the decimal fractions 0.8 + 0.5 + 0.7 = 2.0. Then add the two sums to get 16.)

2. **Guess That Shape:** This shape has three sides that are all equal in length. Guess that shape.
   (You might want to review the names of triangles. The answer is equilateral triangle.)

Tuesday

1. **Calculator Math:** Calculate the following: $\dfrac{2.3 \times 10^{12}}{(1.07 \times 10^5) \times (8.9 \times 10^4)}$.
   (You might want to tell your students how to use scientific notation on their calculators. You can perform the following keystrokes: 2.3 EE 12 / 1.07 EE 5 = / 8.9 EE 4 =. The answer is 241.521.)

2. **Guess That Shape:** This shape has four sides. These sides don't need to have any special relationships. Guess that shape.
   (You might want to review the naming of four-sided polygons. The answer here is quadrilateral.)

Wednesday

1. **Estimation:** 19% of what number is 162? Estimate this number.
   (You might want to tell your students that this is approximately the same as "$\frac{1}{5}$ of what number is 160?" Thus the answer is close to 160 x 5 = 800.)

2. **Guess That Shape:** This shape has four sides. Two adjacent sides are equal to each other, and the other two sides are equal to themselves also (but not necessarily to the other two). This shape is often used in the design of flying toys. Guess that shape.
   (You might want to review the naming of four-sided polygons. The answer is kite.)

Thursday

1. **Calculator Math:** 19% of what number is 162?
   (You might want to tell your students to think of it in the following way: $\frac{19}{100}$ x ? = 162, and thus ? = 162 x $\frac{100}{19}$. Hence, you can perform the following keystrokes to get the answer: 162 x 100 / 19 =. The answer is 852.632.)

2. **Guess That Shape:** This three-sided shape has one right angle. Also, two of its sides are equal in length. Guess that shape.
   (You might want to review naming triangles. The answer is isosceles right triangle.)

TEACHER PAGE: QUARTER 2: WEEK 9

Friday

1. **Mental Math:** Multiply: $\frac{7}{15}$ x 60.

   (You might want to tell your students that in this case it is easier to see the problem as $7 \times \frac{1}{15} \times 60$, and this is $7 \times 4 = 28$.)

2. **Guess That Shape:** This five-sided shape has a very famous model in our nation's capital. It should be noted that the sides are equal to each other. Guess that shape.
   (You should review naming five-sided polygons. The answer is regular pentagon. The term *regular* in this case refers to the fact that all the sides are equal.)

Challenge Problems

1. **Mental Math:** Add the numbers 846 + 375 by thinking about the problem from left to right.
   (You would want to think of it as $800 + 300 = 1,100$; $40 + 70 = 110$; and finally $6 + 5 = 11$. Thus you have $1,100 + 110 + 11 = 1,221$.)

2. **Mental Math:** Multiply the two numbers 368 x 6 by thinking about the problem from left to right.
   (You would want to use the distributive law to see 368 x 6 as $(300 + 60 + 8) \times 6$, which is $300 \times 6 = 1,800$; $60 \times 6 = 360$; and $8 \times 6 = 48$. Thus you have the answer $1,800 + 360 + 48 = 2,208$.)

3. **Mental Math:** Add the following mixed numbers: $\left(3\frac{2}{7} - 5\frac{3}{8}\right) + 4\frac{5}{7}$.

   (You would want to group the compatible numbers together and add them and then subtract the other number. When you do that, you get $\left(3\frac{2}{7} + 4\frac{5}{7}\right) - 5\frac{3}{8} = 8 - 5\frac{3}{8} = 2\frac{5}{8}$.)

4. **Guess That Shape:** This shape has three sides that are all different in length. Guess that shape.
   (You might want to review the different names for triangles. The answer here is scalene triangle.)

5. **Guess That Shape:** This shape has four sides with opposite sides being parallel. The angles are all the same, but the sides are not all the same length. Guess that shape.
   (You might want to review the names and properties of four-sided polygons. This shape is a special kind of parallelogram. With all the angles the same, they must all be 90 degrees. But the sides are not all the same. Hence this shape is a rectangle.)

STUDENT PAGE: QUARTER 2: WEEK 9   **Student**

The theme for week 9 is guess that shape. All of the #2 problems are based on this theme. For each of these problems, you will be given a set of clues. You will then have to become a mathematical detective to figure out the most descriptive name for the plane polygonal shape.

Monday   Name: _____

1. **Mental Math:** Add the following decimals: 5.8 + 3.5 + 6.7.   _____

2. **Guess That Shape:** This shape has three sides that are all equal in length. Guess that shape.   _____

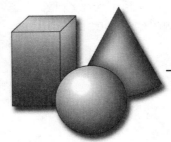

Tuesday   Name: _____

1. **Calculator Math:** Calculate the following: _____
    $$\frac{2.3 \times 10^{12}}{(1.07 \times 10^5) \times (8.9 \times 10^4)}.$$

2. **Guess That Shape:** This shape has four sides. These sides don't need to have any special relationships. Guess that shape.   _____

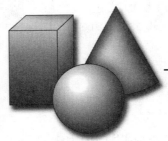

Wednesday   Name: _____

1. **Estimation:** 19% of what number is 162? Estimate this number.   _____

2. **Guess That Shape:** This shape has four sides. Two adjacent sides are equal to each other, and the other two sides are equal to themselves also (but not necessarily to the other two). This shape is often used in the design of flying toys. Guess that shape.   _____

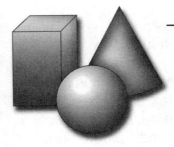

STUDENT PAGE: QUARTER 2: WEEK 9

Thursday

Name: _____

1. **Calculator Math:** 19% of what number is 162?

_____

2. **Guess That Shape:** This three-sided shape has one right angle. Also, two of its sides are equal in length. Guess that shape.

_____

Friday

Name: _____

1. **Mental Math:** Multiply: $\frac{7}{15}$ x 60.

_____

2. **Guess That Shape:** This five-sided shape has a very famous model in our nation's capital. It should be noted that the sides are equal to each other. Guess that shape.

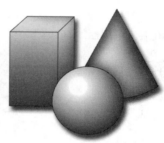

_____

Challenge Problems

Name: _____

1. **Mental Math:** Add the numbers 846 + 375 by thinking about the problem from left to right.

_____

2. **Mental Math:** Multiply the two numbers 368 x 6 by thinking about the problem from left to right.

_____

3. **Mental Math:** Add the following mixed numbers: $3\frac{2}{7}$ - $5\frac{3}{8}$ + $4\frac{5}{7}$.

_____

4. **Guess That Shape:** This shape has three sides that are all different in length. Guess that shape.

_____

5. **Guess That Shape:** This shape has four sides with opposite sides being parallel. The angles are all the same, but the sides are not all the same length. Guess that shape.

_____

TEACHER PAGE: QUARTER 3: WEEK 1

The theme for week 1 is mathematics you might encounter at the mall. All of the #2 problems are based on this theme.

Monday

1. **Mental Math:** Add the following numbers: 56 + 72 + 34 + 13.
    (You might mention that you can sometimes group certain compatible numbers. In this case, 56 + 34 = 90 and 72 + 13 = 85. Thus our answer is 90 + 85 = 175.)

2. **Mall Math:** If the sale price of an item is $24 and it was marked down $33\frac{1}{3}$% from its original price, then what was the original price?
    (You might want to mention that $33\frac{1}{3}$% is probably best dealt with as the fraction $\frac{1}{3}$. You can set this problem up by thinking original less $\frac{1}{3}$ of original equals $24. Thus $\frac{2}{3}$ original equals $24. The answer here is $36.)

Tuesday

1. **Calculator Math:** Add the following list of numbers:
    472 + 825 + 976 + 1063 + 1498 + 593 + 2307.
    (You might have the students start at a common signal and give them a time limit. Also you might want to tell them the appropriate way to add a list of numbers with their calculators. You can perform the following keystrokes: 472 + 825 + 976 + 1063 + 1498 + 593 + 2307 =. The answer is 7734.)

2. **Mall Math:** Every fourth customer at one of the food court restaurants gets a free drink between 2:00 and 3:00 p.m. If there were 76 customers in that time period, how many free drinks were given away?
    (You might want to review proportions. In this problem, you could set up the proportion $\frac{1}{4} = \frac{?}{76}$, and solve for the answer of 19 customers.)

Wednesday

1. **Estimation:** Estimate the following calculation by rounding each number to the nearest tens:
    (38 x 23 x 54)/(76 x 47).
    (You might mention order of operations and the parentheses in this problem. You would have (40 x 20 x 50) / (80 x 50) = (40 x 20) / 80 = 800 / 80. The approximate answer is 10.)

2. **Mall Math:** Which is a better buy: 16 ounces for $0.89 or 20 ounces for $1.09?
    (You might want to talk about the difference between dollars per ounce and ounces per dollar. When you calculate dollars per ounce, the first one becomes $0.89/16 = 0.055625 and the second one becomes $1.09/20 = 0.0545 dollars per ounce. The second is a better buy.)

TEACHER PAGE: QUARTER 3: WEEK 1

Thursday

1. **Calculator Math:** Perform the indicated operations: (38 x 23 x 54)/(76 x 47).
   (You might want to tell students the appropriate way to multiply and divide with the calculator without using memory or writing down intermediate answers. You can perform the following keystrokes: 38 x 23 x 54 = / 76 = / 47 =. Here the answer is 13.212765957.)

2. **Mall Math:** A vacant storefront in the mall is in the shape of a rectangle 20 feet by 40 feet. If it costs $30 per square foot to get a store ready to open, how much will the cost be for this storefront?
   (You might want to review the area of a rectangle. The area of the storefront is 800 square feet and so the cost will be 800 x $30 = $24,000 to get the store ready to open.)

Friday

1. **Mental Math:** Multiply the following numbers: 5 x 79 x 20.
   (You might mention that you can sometimes group certain compatible numbers to make the mental multiplication possible. In this problem, group the 5 x 20 = 100, and hence our answer is 79 x 100 = 7900.)

2. **Mall Math:** The mall owner wants at least 80% of the stores in the mall to be operating. If there are 95 storefronts in the mall, then at least how many must be operating to satisfy the owner?
   (You might want to review the meaning of percent. Here our answer is 95 x 0.80 = 76.)

Challenge Problems

1. **Mall Math:** If it normally takes you 24 minutes to walk a mile in the mall and one loop takes you 10 minutes, then how many loops of the mall will be required to walk a total of two-and-a-half miles?
   (You can attack this problem in two different ways. One way is to figure out how far one loop of the mall is $\frac{10}{24} = \frac{5}{12}$ miles and then divide that into $2\frac{1}{2}$ miles to get the answer of 6 loops. The second way is to figure out how long it would take to walk the total mileage 24 minutes per mile x 2.5 miles = 60 minutes. Then it is easy to see that it will take only 6 loops of 10 minutes each to make the 60 minutes.)

2. **Mall Math:** An item at one store that originally cost $100 was marked down once 50%, and now that sale price is being marked down an additional 25%. What is the new sale price of the item?
   (The first sale price is easy to figure: 50% off $100 is $50, and so the first sale price is $100 - $50 = $50. Now the new sale price is 25% off that $50. The markdown is 25% of $50 = $12.50, and so the new sale price is $50 - $12.50 = $37.50. You might ask your students why the answer isn't just 50% + 25% = 75% off the original $100.)

STUDENT PAGE: QUARTER 3: WEEK 1

The theme for week 1 is mathematics you might encounter at the mall. All of the #2 problems are based on this theme.

Monday   Name: _____

1. **Mental Math:** Add the following numbers: 56 + 72 + 34 + 13.   _____

2. **Mall Math:** If the sale price of an item is $24 and it was marked   _____
   down $33\frac{1}{3}$% from its original price, then
   what was the original price?

Tuesday   Name: _____

1. **Calculator Math:** Add the following list of numbers: _____
   472 + 825 + 976 + 1063 + 1498 + 593 + 2307.

2. **Mall Math:** Every fourth customer at one of the food court res-   _____
   taurants gets a free drink between 2:00 and 3:00 p.m. If there
   were 76 customers in that time pe-
   riod, how many free drinks were
   given away?

Wednesday   Name: _____

1. **Estimation:** Estimate the following calculation by rounding each   _____
   number to the nearest tens: (38 x 23 x 54)/(76 x 47).

2. **Mall Math:** Which is a better buy: 16 ounces for $0.89 or 20   _____
   ounces for $1.09?

STUDENT PAGE: QUARTER 3: WEEK 1

Thursday   Name: _____

1. **Calculator Math:** Perform the indicated operations: _____
   (38 x 23 x 54)/(76 x 47).

2. **Mall Math:** A vacant storefront in the mall is in the shape of a   _____
   rectangle 20 feet by 40 feet. If it costs $30 per square foot to get
   a store ready to open, how much will
   the cost be for this storefront?

Friday   Name: _____

1. **Mental Math:** Multiply the following numbers: 5 x 79 x 20.   _____

2. **Mall Math:** The mall owner wants at least 80% of the stores in   _____
   the mall to be operating. If there are 95 storefronts in the mall,
   then at least how many must be op-
   erating to satisfy the owner?

Challenge Problems   Name: _____

1. **Mall Math:** If it normally takes you 24 minutes to walk a mile in   _____
   the mall and one loop takes you 10 minutes, then how many
   loops of the mall will be required to walk a total of two-and-a-half
   miles?

2. **Mall Math:** An item at one store that originally cost $100 was   _____
   marked down once 50%, and now
   that sale price is being marked
   down an additional 25%. What is
   the new sale price of the item?

76

TEACHER PAGE: QUARTER 3: WEEK 2

The theme for week 2 is running mathematics. All of the #2 problems are based on this theme.

Monday

1. **Mental Math:** Add the following numbers: 36 + 85 + 68.
    (You might want to mention that you can still attack this addition problem mentally even though none of the numbers are compatible. You might want to advise your students to think in the following way: 30 + 80 + 60 = 170; 6 + 5 + 8 = 19; thus you have 170 + 19 = 189 as your answer.)

2. **Running Math:** Many distance running events are measured in kilometers (km). For example, a common road race distance is 5 km (or 5,000 meters). If 1 meter is 3.28 feet, then how many miles are in 5 km?
    (You might want to tell your students to first find the number of feet in 5,000 meters, and then the number of miles in that many feet. You would get 5,000 x 3.28 = 16,400 feet, and thus 16,400 / 5,280 = 3.1 miles. The answer is 3.1 miles.)

Tuesday

1. **Calculator Math:** Calculate the fourth root of 25 using only the square root key on your calculator.
    (You might want to review the meaning of a fourth root, e.g., the fourth root of 16 is 2. Then you might want to relate it to square root. You can perform the following keystrokes: $\sqrt{(\sqrt{25})}$ =. The answer is approximately 2.236.)

2. **Running Math:** A very good sprinter runs 100 yards in 10 seconds. What is that speed in miles per hour?
    (You might want to show them how 100 yds./10 sec. = 600 yds./60 sec. = 600 yds./1 min. = 36,000 yds./60 min. = 36,000 yds./1 hr. and then figure out how many miles in 36,000 yds. Remember, there are 1,760 yards in a mile, so there are 36,000/1,760 = 20.45 miles in 36,000 yards. The answer is 20.45 mph.)

Wednesday

1. **Estimation:** Estimate the following sum: $2.68 + $7.37 + $0.86 + $4.15 + $11.72.
    (You might want to suggest they round the numbers to the nearest dollar or you might want to have them do something like the following: 2 + 7 + 0 + 4 + 11 = 24 and also 1 + 0 + 1 + 0 + 1 = 3, giving $27.)

2. **Running Math:** In distance running, the rate minutes per mile (called the pace) is often used. For example, if I can run at the rate of 6 minutes per mile, then I can run 3 miles in 18 minutes. If a very good runner runs at a 5 minutes-per-mile pace, then how long will it take him or her to run 3.1 miles?
    (You might want to tell the students that this kind of rate multiplied by the number of miles run will give you the time it takes to run. The answer here is 3.1 x 5 = 15.5 minutes or 15 minutes 30 seconds.)

TEACHER PAGE: QUARTER 3: WEEK 2

Thursday

1. **Calculator Math:** Calculate the following sum: $2.68 + $7.37 + $0.86 + $4.15 + $11.72.
   (You might want to start everyone at a common signal and have a time limit for this exercise. You can perform the following keystrokes: 268 + 737 + 86 + 415 + 1172 = and get 2678. It is easy to mentally place the decimal point two places from the right. The answer is $26.78.)

2. **Running Math:** The marathon is the longest traditional running event. It is 26.2 miles long. What pace must a runner run to break 3 hours in the marathon?
   (You might want to mention that this is simply a division problem instead of multiplication as in #2 for Wednesday. You might first change 3 hours to 180 minutes and then divide 180 by 26.2 to get 6.87 minutes per mile. The answer is 6.87 minutes per mile or 6 minutes 52 seconds per mile.)

Friday

1. **Mental Math:** Multiply the following numbers: 60 x 15.
   (You might want to instruct the students in looking at the above problem in the following way: 60 x 15 = (6 x 15) x 10 = 90 x 10 = 900.)

2. **Running Math:** The world record for the marathon at the beginning of the year 2000 was 2 hours, 5 minutes, and 42 seconds. What was the pace of that runner?
   (You might want to mention that this is simply a division problem similar to problem #2 for Thursday. What makes this more challenging is the need to convert the time to minutes before you divide. For 2 hours, 5 minutes, and 42 seconds, you get 120 + 5 + 42 / 60 = 125.7 minutes. Thus 125.7 / 26.2 results in 4.7977 minutes per mile. The answer is 4 minutes 47 seconds per mile.)

Challenge Problems

1. **Sports Math:** The world record for women in 10,000 meters in track is 29 minutes 31.78 seconds. Use the fact that 10,000 meters is approximately 6.2 miles to figure how long it took this runner to run each mile.
   (You will want to change the time in minutes and seconds into just minutes. So the 29 minutes 31.78 seconds becomes 29.53 minutes. Next, divide this time by the 6.2 miles to get 29.53 / 6.2 = 4.76 minutes, or approximately 4 minutes 46 seconds per mile.)

2. **Sports Math:** The world record for men in the high jump is 2.45 meters. Use the fact that 1 inch is approximately 2.54 centimeters to figure how many feet in 2.45 meters.
   (You will want to convert the meters to centimeters by simply moving the decimal point two places to the right to become 245 centimeters. Now divide 245 centimeters by 2.54 to get the number of inches, 245 / 2.54 = 96.5 inches. Finally, convert this to feet and inches by dividing by 12, to get 8 feet $\frac{1}{2}$ inch.)

STUDENT PAGE: QUARTER 3: WEEK 2

Student

The theme for week 2 is running mathematics. All of the #2 problems are based on this theme.

- -

Monday   Name:_____

1. **Mental Math:** Add the following numbers: 36 + 85 + 68.   _____

2. **Running Math:** Many distance running events are measured in   _____
    kilometers (km). For example, a common road race distance is
    5 km (or 5,000 meters). If 1 meter is 3.28 feet,
    then how many miles are in 5 km?

- -

Tuesday   Name:_____

1. **Calculator Math:** Calculate the fourth root of 25 using only the   _____
    square root key on your calculator.

2. **Running Math:** A very good sprinter runs 100   _____
    yards in 10 seconds. What is that speed in
    miles per hour?

- -

Wednesday   Name:_____

1. **Estimation:** Estimate the following sum: _____
    $2.68 + $7.37 + $0.86 + $4.15 + $11.72.

2. **Running Math:** In distance running, the rate minutes per mile   _____
    (called the pace) is often used. For example, if I can run at the
    rate of 6 minutes per mile, then I can run 3
    miles in 18 minutes. If a very good runner
    runs at a 5 minutes-per-mile pace, then
    how long will it take him or her to run 3.1
    miles?

STUDENT PAGE: QUARTER 3: WEEK 2

Student

Thursday

Name: _____

1. **Calculator Math:** Calculate the following sum: _____
   $2.68 + $7.37 + $0.86 + $4.15 + $11.72.

2. **Running Math:** The marathon is the longest traditional running _____
   event. It is 26.2 miles long. What pace must a
   runner run to break 3 hours in the marathon?

Friday

Name: _____

1. **Mental Math:** Multiply the following numbers: 60 x 15. _____

2. **Running Math:** The world record for the marathon at the begin- _____
   ning of the year 2000 was 2 hours, 5 minutes,
   and 42 seconds. What was the pace of that
   runner?

Challenge Problems

Name: _____

1. **Sports Math:** The world record for women in 10,000 meters in _____
   track is 29 minutes 31.78 seconds. Use the
   fact that 10,000 meters is approximately 6.2
   miles to figure how long it took this runner
   to run each mile.

2. **Sports Math:** The world record for men in the high jump is 2.45 _____
   meters. Use the fact that 1 inch is approximately 2.54 centimeters to figure how many feet in 2.45 meters.

TEACHER PAGE: QUARTER 3: WEEK 3

The theme for this week is North American animals. All of the #2 problems are based on this theme.

Monday

1. **Mental Math:** Calculate 17% of 300.
   (You might want to remind your students that the % symbol is short for /100. Thus 17% is the same as 17/100 x 300 = 17 x 1/100 x 300 = 17 x 3 = 51.)

2. **North American Animals:** A grey wolf or timber wolf may weigh as much as 175 pounds and be over 6 feet long (including the tail). If 1 ounce = 29.5 grams, then how many kilograms (kg) does this timber wolf weigh?
   (You might want to tell your students to convert pounds to ounces 175 x 16 = 2,800 and then multiply this by 29.5 grams per ounce to get 2,800 x 29.5 = 82,600 grams. Divide by 1,000 to get 82.6 kg.)

Tuesday

1. **Calculator Math:** Calculate the "special" product 7 x 6 x 5 x 4 x 3 x 2 x 1. This is abbreviated by 7! or "7 factorial." See if your calculator has a ! key on it. Try this problem both ways if it does.
   (You might want to help your students look for the ! key on their calculators or show them one that does have it. If the calculator has the factorial key (!), then you would simply perform the keystrokes: 7 ! =. The answer here is 5,040.)

2. **North American Animals:** A Kodiak brown bear might weigh as much as 780 kilograms (kg). If 1 ounce = 29.5 grams, then how many pounds does this Kodiak brown bear weigh?
   (You might want to first have your students change the 780 kg to 780,000 g and then figure out how many ounces in this number. First you would get 780,000 / 29.5 = 26,440.68 ounces, and then 26,440.68 / 16 = 1,653 pounds.)

Wednesday

1. **Estimation:** Estimate 42.9% of 618.
   (You might want to tell your students that there are several good approaches to this estimation problem. One way would be to think of it as $\frac{40}{100}$ of 600 and come up with 240 as your answer.)

2. **North American Animals:** The bison or buffalo population in North America shrank to as few as 1,000 around 1890. Today there are around 200,000 buffalo on farms and ranches in North America. What percent increase in the buffalo population occurred between 1890 and 2000?
   (You might want to review the meaning of percent increase. Then since there was an increase of 199,000 in the time frame, the percent increase would be (199,000 / 1,000) x 100 = 19,900%.)

81

TEACHER PAGE: QUARTER 3: WEEK 3

Thursday

1. **Calculator Math:** Calculate 43.9% of 618 without using the % key on your calculator (if it has one).
   (You might want to review the meaning of percent. In this case, simply multiply 0.439 x 618 to get your answer of 271.302.)

2. **North American Animals:** A North American coyote can run with speeds up to 40 miles per hour (or 59 feet per second). How long would it take a coyote to run a fourth of a mile (one lap around a track)?
   (You might want to tell your students to use the rate of 59 ft./sec. and to convert the fourth of a mile to feet ($\frac{1}{4}$ of 5,280 = 1,320). Then they can use Distance = Rate x Time to find Time = 1,320/ 59 = 22.37 seconds.)

Friday

1. **Mental Math:** Can you figure out the unknown number in the equation 2 x __ + 50 = 140?
   (You might want to go through a concrete substitution for the unknown to show the students what arithmetic operations are performed in this equation. The answer is 45.)

2. **North American Animals:** The world has a horse population of approximately 61 million. It is estimated that around 100,000 of the horses in the United States are miniature horses. What percent (of the world horse population) do these 100,000 horses represent?
   (You might want to review the meaning of percent. The answer here is simply (100,000 / 61,000,000) x 100 = 0.16%.)

Challenge Problems

1. **Calculator Math:** Here is a way to approximate a square root without using that key on your calculator. If we want to approximate the square root of 3:
   1) We will start with a guess of 2, then calculate 3 / 2 = 1.5. If that number is the same as your guess, then that is the answer. If not, then average these two numbers, i.e., take (2 + 1.5) / 2 = to get 1.75. This answer (1.75) should be "better" than the 2 you started with.
   2) Now repeat step 1 using 1.75 as your guess.
   3) Now repeat step 1 using the answer you got in step 2.
   4) Continue as above until your answer doesn't change.
   (At step 2 you compute (1.75 + 3 / 1.75) / 2 = and get 1.73214 as your answer. At step 3 you compute (1.73214 + 3 / 1.73214) / 2 = and get 1.73205. At step 4 you compute (1.73205 + 3 / 1.73205) / 2 = and get 1.73205. This is the same as in step 3, so this is the approximate square root for 3.)

STUDENT PAGE: QUARTER 3: WEEK 3

Student

The theme for this week is North American animals. All of the #2 problems are based on this theme.

Monday   Name: _____

1. **Mental Math:** Calculate 17% of 300.   _____

2. **North American Animals:** A grey wolf or
   timber wolf may weigh as much as 175
   pounds and be over 6 feet long (including the tail). If 1 ounce = 29.5 grams,
   then how many kilograms (kg) does
   this timber wolf weigh?

Tuesday   Name: _____

1. **Calculator Math:** Calculate the "special" product 7 x 6 x 5 x 4 x   _____
   3 x 2 x 1. This is abbreviated by 7! or "7 factorial." See if your
   calculator has a ! key on it. Try this problem both ways if it does.

2. **North American Animals:** A Kodiak brown
   bear might weigh as much as 780 kilograms
   (kg). If 1 ounce = 29.5 grams, then how
   many pounds does this Kodiak brown
   bear weigh?

Wednesday   Name: _____

1. **Estimation:** Estimate 42.9% of 618.   _____

2. **North American Animals:** The bison or buffalo population in North America shrank to
   as few as 1,000 around 1890. Today there
   are around 200,000 buffalo on farms and
   ranches in North America. What percent increase in the buffalo population occurred
   between 1890 and 2000?

STUDENT PAGE: QUARTER 3: WEEK 3

Student

Thursday   Name: _____

1. **Calculator Math:** Calculate 43.9% of 618 without using the %   _____
   key on your calculator (if it has one).

2. **North American Animals:** A North American
   coyote can run with speeds up to 40 miles
   per hour (or 59 feet per second). How long   _____
   would it take a coyote to run a fourth of a
   mile (one lap around a track)?

Friday   Name: _____

1. **Mental Math:** Can you figure out the unknown number in the   _____
   equation 2 x __ + 50 = 140?

2. **North American Animals:** The world has a
   horse population of approximately 61 million.
   It is estimated that around 100,000 of the
   horses in the United States are miniature   _____
   horses. What percent (of the world horse
   population) do these 100,000 horses represent?

Challenge Problems   Name: _____

1. **Calculator Math:** Here is a way to approximate a square root   _____
   without using that key on your calculator. If we want to approximate the square root of 3:
   1) We will start with a guess of 2, then calculate 3 / 2 = 1.5. If
      that number is the same as your guess, then that is the an-
      swer. If not, then average these two numbers, i.e., take (2 +
      1.5) / 2 = to get 1.75. This answer (1.75) should be "better"
      than the 2 you started with.
   2) Now repeat step 1 using 1.75 as your guess.
   3) Now repeat step 1 using the answer you got in step 2.
   4) Continue as above until your answer doesn't change

TEACHER PAGE: QUARTER 3: WEEK 4

The theme for week 4 is American history puzzles. All of the #2 problems for this week will be posed as puzzles about famous events in American history.

Monday

1. **Mental Math:** Find $33\frac{1}{3}$% of $12,000.
   (You might want to explain how some percents can be thought of more easily in terms of fractions. In our case, $33\frac{1}{3}$% is the same as the fraction $\frac{1}{3}$ and so our problem is the same as finding $\frac{1}{3}$ of $12,000, which is $4,000.)

2. **American History Puzzle:** The famous Boston Tea Party happened in what year? Hint! The sum of the digits is 18 and the hundreds digit is equal to the tens digit.
   (You might want to review problem-solving strategies and setting up equations. In this problem the equations are $th + h + te + u = 18$ and $h = te$. You might guess that th = 1 and h = 7, and then you would be able to deduce that te = 7 and u = 3. The answer is 1773.)

Tuesday

1. **Calculator Math:** Find the area of a circular table whose diameter is 10 feet.
   (You might want to review the formula for the area of a circle and how to use the π key on a calculator. Since the formula for the area of a circle is $A = r^2 \times \pi$ and r = 5 feet, you can perform the following keystrokes: 5 $x^2 \times \pi =$ to get the answer 78.5398 square feet.)

2. **American History Puzzle:** The United States was drawn into World War II when Pearl Harbor was bombed in this year. Hint! Consider the number formed from the tens and units digits along with the number formed from the thousands and hundreds digits. Their sum is 60 and their difference is 22.
   (You might want to review problem-solving strategies and setting up equations. In this case the equations would be $f + s = 60$ and $s - f = 22$. You should know that f = 19, so it is easy to see that s = 41. The answer is 1941.)

Wednesday

1. **Estimation:** Estimate how much tax you will have to pay at the restaurant if you bought items for $2.95, $3.95, $1.79, $1.09, and $1.09 with a tax rate of 8%.
   (You might want to mention rounding to the nearest "nice" number before you sum the items and then add them and multiply by 0.08. Thus you get $3 + $4 + $2 + $1 + $1 = $11, and 8% of that is $0.88.)

2. **American History Puzzle:** Thomas Jefferson became president in what year? Hint! The square of the number is 3,243,601.
   (You might want to review the concept of square and square root. Since $year^2$ = 3,243,601, then $year$ is the square root of 3,243,601, which is 1801. The answer is 1801.)

85

TEACHER PAGE: QUARTER 3: WEEK 4 Teacher

Thursday

1. **Calculator Math:** Calculate $(1.03)^{16}$ by using only the square function of your calculator. (You might want to review what it means to square something. Since the square of the square is the same as the fourth power, then the sixteenth power is the same as the square of the square of the square of the square. You can perform the following keystrokes: 1.03 x^2 x^2 x^2 x^2. The answer is 1.604706439.)

2. **American History Puzzle:** In what year did U.S. troops pull out of the war in Vietnam? Hint! Add 37 to the year whose square root is 44. (You might want to review translating words to numbers and number operations. In this case you get $year = 44 \times 44 + 37 = 1973$.)

Friday

1. **Mental Math:** Add 2.3 + 4.2 + 7.5. (You might want to mention that in this case, you can find compatible numbers in the decimals .3 + .2 + .5 = 1. Hence the sum is 2 + 4 + 7 + 1 = 14.)

2. **American History Puzzle:** In what year did man first walk on the surface of the moon? Hint! If you take the number formed from the tens digit and the units digit, add 1 to it, divide that by 7, then multiply the result by 5, the answer is 50. (You might want to discuss how to "undo" a string of arithmetic operations by reversing the order and applying the "opposite" operation. Thus you take 50, divide it by 5 to get 10, multiply that by 7 to get 70, and finally subtract 1 to get the number 69. Hence the answer is 1969.)

Challenge Problems

1. **American History Puzzle:** Wilbur and Orville Wright made their first successful powered flight on December 17 of the year whose sum of digits is 13 and whose units digit divides into the hundreds digit evenly. What is the year? (You can guess that the year is either in the late 1800s or early 1900s. But with the two clues you will be able to figure out that it can't be in the 1800s and is either 1903 or 1921. Hopefully, you will realize that 1921 is too late, and so the answer must be 1903.)

2. **American Presidents:** Dwight Eisenhower was born in Kansas, and went on to become a celebrated general during World War II. Before he became President of the United States he was president of Columbia University in New York. Eisenhower died in 1969. He became President of the United States in the year whose sum of tens and units is 8. Also, the tens digit minus the units digit is 2. What is the year? (You should have no problem with the fact that the year is in the 1900s. So you only have to figure out the tens and units digits. With the two clues, it should be an easy matter to settle on 5 and 3. Thus the year is 1953.)

STUDENT PAGE: QUARTER 3: WEEK 4

The theme for week 4 is American history puzzles. All of the #2 problems for this week will be posed as puzzles about famous events in American history.

Monday   Name: _____

1. **Mental Math:** Find $33\frac{1}{3}\%$ of $12,000.   _____

2. **American History Puzzle:** The famous Boston Tea   _____
    Party happened in what year? Hint! The
    sum of the digits is 18 and the hun-
    dreds digit is equal to the tens digit.

Tuesday   Name: _____

1. **Calculator Math:** Find the area of a circular table whose diam-   _____
    eter is 10 feet.

2. **American History Puzzle:** The United States was drawn into   _____
    World War II when Pearl Harbor was bombed in this year. Hint!
    Consider the number formed from the tens and units digits along
    with the number formed from the thou-
    sands and hundreds digits. Their sum is
    60 and their difference is 22.

Wednesday   Name: _____

1. **Estimation:** Estimate how much tax you will have to pay at the   _____
    restaurant if you bought items for $2.95, $3.95, $1.79, $1.09,
    and $1.09 with a tax rate of 8%.

2. **American History Puzzle:** Thomas Jefferson be-   _____
    came president in what year? Hint! The square
    of the number is 3,243,601.

STUDENT PAGE: QUARTER 3: WEEK 4 ### Student

Thursday   **Name:** _____

1. **Calculator Math:** Calculate $(1.03)^{16}$ by using only the square   _____
   function of your calculator.

2. **American History Puzzle:** In what year did U.S. troops    _____
   pull out of the war in Vietnam? Hint! Add
   37 to the year whose square
   root is 44.

Friday   **Name:** _____

1. **Mental Math:** Add 2.3 + 4.2 + 7.5.   _____

2. **American History Puzzle:** In what year did      _____
   man first walk on the surface of the moon? Hint!
   If you take the number formed from the tens
   digit and the units digit, add 1 to it, divide that
   by 7, then multiply the result by 5, the answer
   is 50.

Challenge Problems   **Name:** _____

1. **American History Puzzle:** Wilbur and Orville Wright made their   _____
   first successful powered flight on December 17 of the year whose
   sum of digits is 13 and
   whose units digit divides
   into the hundreds digit
   evenly. What is the year?

2. **American Presidents:** Dwight Eisenhower was born in Kan-   _____
   sas, and went on to become a celebrated general during World
   War II. Before he became President of the United States he was
   president of Columbia University in New York. Eisenhower died
   in 1969. He became President of the United States in the year
   whose sum of tens and units is 8. Also, the tens digit minus the
   units digit is 2. What is the year?

TEACHER PAGE: QUARTER 3: WEEK 5

The theme for week 5 is famous seventeenth-century mathematicians. All the #2 problems are based on this theme.

Monday

1. **Mental Math:** Add 5.63 + 2.97.
   (You might want to mention that if you add something to one number and subtract the same something from the other number, then the answer remains the same. For example, add 0.03 to 2.97 and subtract 0.03 from 5.63 to give us 5.63 + 2.97 = 5.60 + 3 = 8.60.)

2. **Seventeenth-Century Mathematicians:** Blaise Pascal was a French mathematician who created a "triangle" of numbers that will help you solve the following problem. A pizza maker has 4 different ingredients that he can use to make a pizza. If he can put either 0, 1, 2, 3, or 4 ingredients on a pizza (no half-and-half pizzas), count the number of different pizzas he can make. (You might want to clarify what is being asked here, and you might want to show the students how to make Pascal's triangle. You can create Pascal's triangle with 1 at the top, then 1   1 in the row under that, 1   2   1 in the row under that, 1   3   3   1, in the row under that and so on. It will look like

$$1$$
$$1 \qquad 1$$
$$1 \qquad 2 \qquad 1$$
$$1 \qquad 3 \qquad 3 \qquad 1.$$

   In the next row the numbers would be

$$1 \qquad 4 \qquad 6 \qquad 4 \qquad 1$$

   each of those being the sum of the ones in the row above on either side. This row can be used to come up with the answer 1 + 4 + 6 + 4 + 1 = 16 different pizzas can be made.)

Tuesday

1. **Calculator Math:** Find the circumference of a circular table with a diameter of 10 feet.
   (You might want to review what the circumference of a circle is and the formula for finding it. Also you might want to show your students how to use their π key. Since the C = D x π, you can perform the following keystrokes: 10 x π =. The answer is 31.41592654 feet.)

2. **Seventeenth-Century Mathematicians:** Pierre de Fermat was a French mathematician who liked to write his results in the margins of books. He discovered that any prime number with a remainder of 1, when divided by 4, can be written as the sum of two squares, e.g., 29 = 25 + 4. Express each of the numbers 53 and 73 as the sum of two squares.
   (You might want to review the concepts of prime number and square number. By using the guess-and-check method, you should find 53 = 49 + 4 and 73 = 64 + 9.)

TEACHER PAGE: QUARTER 3: WEEK 5 — Teacher

Wednesday

1. **Estimation:** Estimate the following calculation: $\dfrac{8.61 \times 10^{-6}}{(2.77 \times 10^{-5})(1.02 \times 10^{-4})}$.

    (You might want to review scientific notation and how to look at the powers of 10 separately. You would then get

    $$\frac{9}{3 \times 1} \quad \times \quad \frac{10^{-6}}{10^{-5} \times 10^{-4}} = 3 \times \frac{10^{-6}}{10^{-9}} = 3 \times 10^{3}.)$$

2. **Seventeenth-Century Mathematicians:** Rene Descartes was a French mathematician who is generally credited with the discovery of the subject of analytic geometry. He also discovered a formula ($v - e + f = 2$) that relates the number of vertices, edges, and faces of a three-dimensional solid called a polyhedron. A cube is a polyhedron. Count the number of vertices, edges, and faces of a cube. Does Descartes' formula work?
    (You might want to go over vertices, edges, and faces of a polyhedron. The answers are $v = 8$, $e = 12$, and $f = 6$. The formula does work: 8 - 12 + 6 = 2.)

Thursday

1. **Calculator Math:** Perform the following calculation: $\dfrac{8.61 \times 10^{-6}}{(2.77 \times 10^{-5})(1.02 \times 10^{-4})}$.
    (You might want to review scientific notation on the calculator. You can perform the following keystrokes: 8.61 EE (-) 6 / 2.77 EE (-) 5 = / 1.02 EE (-) 4 =. The answer is 3047.356127.)

2. **Seventeenth-Century Mathematicians:** Isaac Newton was an English mathematician who is regarded as one of the greatest mathematicians of all time. He is credited with the discovery of calculus. He did so in what year? Hint! When you add 15 to this year and take the square root, you get 41. (You might want to talk about problem-solving strategies and the value of setting up an equation: $\sqrt{year + 15} = 41$. To solve this equation, you square 41 and subtract 15. The answer is 1666.)

Friday

1. **Mental Math:** Calculate $35 \times \frac{4}{5} - 35 \times \frac{3}{5}$.
    (You might want to tell your students about the distributive property. It can be used here to make this a mental calculation. You get $35 \times (\frac{4}{5} - \frac{3}{5}) = 35 \times \frac{1}{5} = 7$.)

2. **Seventeenth-Century Mathematicians:** Gottfried Leibnitz was a German mathematician who also discovered calculus on his own, but his discovery was after that of Isaac Newton. Leibnitz discovered the quotient and product rules of calculus in the year that is divisible by both 67 and 25. What is this year?
    (You might want to mention what it means to be divisible by a number. Thus year = 67 x 25 x ?, and since 67 x 25 is already 1675, then ? must be 1. The answer here is 1675.)

Challenge Problem

1. Perform the following addition in your head without the use of pencil and paper or calculator: 2 + 7 + 5 + 8 + 4 + 9 + 6 + 4. The answer is (2 + 7 + 5 + 8 + 4 + 9 + 6 + 4 = 45.)

STUDENT PAGE: QUARTER 3: WEEK 5   Student

The theme for week 5 is famous seventeenth-century mathematicians. All the #2 problems are based on this theme.

- -

Monday   Name: _____

1. **Mental Math:** Add 5.63 + 2.97.   _____

2. **Seventeenth-Century Mathematicians:** Blaise Pascal was   _____
   a French mathematician who created a "triangle" of numbers that would help you solve the following problem. A
   pizza maker has 4 different ingredients that he can use
   to make a pizza. If he can put either 0, 1, 2, 3, or 4
   ingredients on a pizza (no half-and-half pizzas), count
   the number of different pizzas he can make.

- -

Tuesday   Name: _____

1. **Calculator Math:** Find the circumference of a circular table with   _____
   a diameter of 10 feet.

2. **Seventeenth-Century Mathematicians:** Pierre de   _____
   Fermat was a French mathematician who liked
   to write his results in the margins of books. He
   discovered that any prime number with a remainder of 1, when divided by 4, can be written as the
   sum of two squares, e.g., 29 = 25 + 4. Express each of
   the numbers 53 and 73 as the sum of two squares.

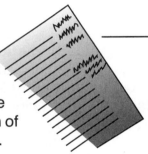

- -

Wednesday   Name: _____

1. **Estimation:** Estimate the following calculation: _____

   $$\frac{8.61 \times 10^{-6}}{(2.77 \times 10^{-5})(1.02 \times 10^{-4})}.$$

2. **Seventeenth-Century Mathematicians:** Rene Descartes was
   a French mathematician who is generally credited with the dis-   _____
   covery of the subject of analytic geometry. He also
   discovered a formula ($v - e + f = 2$) that relates the
   number of vertices, edges, and faces of a three-
   dimensional solid called a polyhedron. A cube is
   a polyhedron. Count the number of vertices,
   edges, and faces of a cube. Does Descartes' formula work?

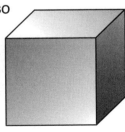

STUDENT PAGE: QUARTER 3: WEEK 5   **Student**

Thursday   Name: _____

1. **Calculator Math:** Perform the following calculation: _____

   $$\frac{8.61 \times 10^{-6}}{(2.77 \times 10^{-5})(1.02 \times 10^{-4})}.$$

2. **Seventeenth-Century Mathematicians:** Isaac New-      _____
   ton was an English mathematician who is regarded
   as one of the greatest mathematicians of all time.
   He is credited with the discovery of calculus. He did
   so in what year? Hint! When you add 15 to this year
   and take the square root, you get 41.

Friday   Name: _____

1. **Mental Math:** Calculate $35 \times \frac{4}{5} - 35 \times \frac{3}{5}$.   _____

2. **Seventeenth-Century Mathematicians:** Gottfried      _____
   Leibnitz was a German mathematician who also dis-
   covered calculus on his own, but his discovery
   was after that of Isaac Newton. Leibnitz discovered the quotient and product rules of calculus in the year that is divisible by both 67
   and 25. What is this year?

Challenge Problem   Name: _____

1. **Mental Math:** Perform the following addi-   _____
   tion in your head without the use of pencil
   and paper or calculator:
   $2 + 7 + 5 + 8 + 4 + 9 + 6 + 4.$

TEACHER PAGE: QUARTER 3: WEEK 6

The theme for week 6 is States Geography Math, the mathematics that will help you learn some interesting facts about the geography of our states. All the #2 problems are based on this theme.

Monday

1. **Mental Math:** How many minutes are there between 9:30 a.m. and 1:00 p.m.?
   (You might want to review clock arithmetic and the conversion of hours to minutes. The answer is $3\frac{1}{2}$ hours = 210 minutes.)

2. **States Geography Math:** In California, the difference in elevation between its highest point (Mount Whitney) and its lowest point (Death Valley) is 14,776 feet. Death Valley is 282 feet below sea level. What is the elevation of Mt. Whitney?
   (You might want to review what it means to be below sea level, and the difference between a positive and a negative number. You should set up ? - (-282) = 14,776. But that is the same as ? + 282 = 14,776, and so ? = 14,776 - 282 = 14,494. The answer is 14,494 feet.)

Tuesday

1. **Calculator Math:** Calculate the average of these six test scores: 92, 88, 94, 86, 99, and 89.
   (You might want to review how to calculate an average and how to use the calculator without using memory or writing down intermediate answers. You can perform the following keystrokes: 92 + 88 + 94 + 86 + 99 + 89 = / 6 = . The average is 91.33.)

2. **States Geography Math:** Rhode Island is the smallest state in land area (1,231 square miles). If it were a square in shape, what would the length of each side of the square be?
   (You might want to review the meaning of area and the formula for the area of a square. You can set up s^2 = 1,231 and solve for s by taking the square root of 1,231. The answer is 35.1 miles.)

Wednesday

1. **Estimation:** If $\frac{1}{3}$ of the 14,637 people at a concert buy a T-shirt that costs $15.00, estimate the total amount spent on T-shirts.
   (You might want to help your students with the words so that they understand they you have approximately $\frac{1}{3}$ of 15,000 = 5,000 T-shirts sold at $15.00 each for an approximate amount of $75,000.)

2. **States Geography Math:** Nevada is the seventh-largest state in land area (110,567 square miles). There are only 16 counties in the state. If each county was the same size and each was shaped like a square, how long would the side of each square be?
   (You might want to review the meaning of area and the formula for the area of a square. Thus each county would have an area of 110,567 / 16 = 6,910.44. Then set up s^2 = 6,910.44 and take the square root of 6,910.44 to get the side. The answer is 83.1 miles on the side of each county.)

Teacher

Thursday

1. **Calculator Math:** If $\frac{1}{3}$ of the 14,637 people at a concert buy a T-shirt that costs $15.00, then calculate how much is spent on T-shirts.
   (You might want to advise your students to use $\frac{1}{3}$ on their calculators intead of 0.333. You can perform the following keystrokes: (1 / 3) x 14637 = x 15 =. The answer is $73,185.00.)

2. **States Geography Math:** New Mexico is the fifth-largest state in land area (121,598 square miles). Its highest point (Wheeler Peak) is at an elevation of 13,161 feet, and its lowest point (Red Bluff Reservoir) is 10,319 feet below that. What is the elevation of Red Bluff Reservoir? (You might want to review the components of a subtraction problem. Thus you set up 13,161 - ? = 10,319 or ? = 13,161 - 10,319. The answer is 2,842 feet.)

Friday

1. **Mental Math:** Perform the subtraction: $9 - 3\frac{4}{7}$.

   (You might tell your students that if you add the same number to both quantities in the problem, the answer remains the same, i.e., add $\frac{3}{7}$ to both numbers in the problem to get $9\frac{3}{7} - 4 = 5\frac{3}{7}$.)

2. **States Geography Math:** The highest point in Louisiana (Driskill Mountain) is one of the lowest high points among the states. In fact, it is lower than many states' lowest points. The sum of the digits in this three-digit number is 13. The hundreds and units digits are the same, and the tens digit is two less. What is the elevation of Driskill Mountain?
   (You might want to review problem-solving strategies and, in particular, the value of setting up equations for this kind of problem. You get the equations $h + t + u = 13$, $h = u$, and $t = h - 2$. With a little guess-and-check, you can figure out that $h = u = 5$ and $t = 3$. The answer is 535 feet.)

Challenge Problem

1. **States Geography Math:** Florida was the 27th state admitted to the United States. Its highest point is only 345 feet in elevation (Britton Hill). This ranks 50th among the states for their highest points. The state ranks third in water area (11,761 square miles). If this water area was a single circular lake, then how big would this lake be?
   (You will need to review the area of a circle ($A = \pi r^2$). So, since you know the area, the equation can be solved for r by dividing 11,761 by π, then taking the square root. The answer is 61.2 miles in radius or 122.4 miles in diameter.)

STUDENT PAGE: QUARTER 3: WEEK 6

The theme for week 6 is States Geography Math, the mathematics that will help you learn some interesting facts about the geography of our states. All the #2 problems are based on this theme.

Monday   Name:_____

1. **Mental Math:** How many minutes are there between 9:30 a.m.   _____
   and 1:00 p.m.?

2. **States Geography Math:** In California, the difference in eleva-   _____
   tion between its highest point (Mount
   Whitney) and its lowest point (Death
   Valley) is 14,776 feet. Death Valley is
   282 feet below sea level. What is the
   elevation of Mt. Whitney?

Tuesday   Name:_____

1. **Calculator Math:** Calculate the average of these six test scores: _____
   92, 88, 94, 86, 99, and 89.

2. **States Geography Math:** Rhode Island is   _____
   the smallest state in land area (1,231 square
   miles). If it were a square in shape, what
   would the length of each side of the square
   be?

Wednesday   Name:_____

1. **Estimation:** If $\frac{1}{3}$ of the 14,637 people at a concert buy a T-shirt   _____
   that costs $15.00, estimate the total amount spent on T-shirts.

2. **States Geography Math:** Nevada is the seventh-largest state   _____
   in land area (110,567 square miles).
   There are only 16 counties in the
   state. If each county was the same
   size and each was shaped like a
   square, how long would the side of
   each square be?

STUDENT PAGE: QUARTER 3: WEEK 6

Student

Thursday   Name: _____

1. **Calculator Math:** If $\frac{1}{3}$ of the 14,637 people at a concert buy a   _____
   T-shirt that costs $15.00, then calculate how much is spent on
   T-shirts.

2. **States Geography Math:** New Mexico is the fifth-largest state   _____
   in land area (121,598 square miles).
   Its highest point (Wheeler Peak) is
   at an elevation of 13,161 feet, and
   its lowest point (Red Bluff Reservoir)
   is 10,319 feet below that. What is the
   elevation of Red Bluff Reservoir?

New Mexico

Friday   Name: _____

1. **Mental Math:** Perform the subtraction: $9 - 3\frac{4}{7}$.   _____

2. **States Geography Math:** The highest point in Louisiana (Driskill   _____
   Mountain) is one of the lowest high points among the states. In
   fact, it is lower than many states' low-
   est points. The sum of the digits in
   this three-digit number is 13. The
   hundreds and units digits are the
   same, and the tens digit is two less.
   What is the elevation of Driskill
   Mountain?

Louisiana

Challenge Problem   Name: _____

1. **States Geography Math:** Florida was the 27th state admitted   _____
   to the United States. Its highest point is only 345 feet in elevation (Britton Hill). This ranks 50th
   among the states for their highest
   points. The state ranks third in water area (11,761 square miles). If this
   water area was a single circular lake,
   then how big would this lake be?

Florida

TEACHER PAGE: QUARTER 3: WEEK 7

The theme for week 7 will be making change. You will be given the amount of a purchase and the amount of money given to pay for the purchase. You are to figure the change to be given back by using the least amount of bills and coins in your cash register. The register contains the usual coins (pennies, nickels, dimes, and quarters) and the usual bills (ones, fives, tens, and twenties). All the #2 problems are based on this theme.

Monday

1. **Mental Math:** Calculate 30% of 620.
   (You might mention the basic mental calculation here is 3 x 620 = 1860. The rest is figuring out where the decimal place goes. By estimation, it should be about $\frac{1}{3}$ of 600, or 200. Therefore the answer is 186.0.)

2. **Make Change:** The purchase is $0.26, and the amount tendered is $1.00.
   (You might suggest that your students try making change mentally by adding up to the amount tendered. Then they could check it with pencil-and-paper methods. The answer is 4-P, 2-D, and 2-Q.)

Tuesday

1. **Calculator Math:** Perform the following calculation: $\dfrac{2 - \sqrt{5}}{\sqrt{3}}$.

   (You might want to talk about order of operations and how to do this without having to use memory or writing down intermediate answers. You can perform the following keystrokes: 2 - √ 5 = / √ 3 =. The answer is -0.136294.)

2. **Make Change:** The purchase is $3.91, and the amount tendered is $10.00.
   (You might suggest that your students try making change mentally by adding up to the amount tendered. Then they could check it with pencil-and-paper methods. The answer is 4-P, 1-N, 1-$1, and 1-$5.)

Wednesday

1. **Estimation:** Gary exercises about 40 minutes every day. How much does he exercise in a year?
   (You might want to tell your students to convert the 40 minutes to $\frac{2}{3}$ of an hour, and then figure $\frac{2}{3}$ of about 360 days. One approximate answer is 240 hours.)

2. **Make Change:** The purchase is $2.19, and the amount tendered is $5.04.
   (You might suggest that your students try making change mentally by adding up to the amount tendered. Then they could check it with pencil-and-paper methods. The answer is 1-D, 3-Q, and 2-$1.)

TEACHER PAGE: QUARTER 3: WEEK 7 | Teacher

Thursday

1. **Calculator Math:** Perform the calculation: $\dfrac{(4.75 - 2.93) \times (12.58 + 4.05)}{(7.29 - 3.49)}$.

   (You might want to talk about how to use the parentheses key on your calculator. You can perform the following keystrokes: (4.75 - 2.93) x (12.58 + 4.05) / (7.29 - 3.49) =. The answer is 7.964894737.)

2. **Make Change:** The purchase is $6.11, and the amount tendered is $11.25.
   (You might suggest that your students try making change mentally by adding up to the amount tendered. Then they could check it with pencil-and-paper methods. The answer is 4-P, 1-D, and 1-$5.)

Friday

1. **Mental Math:** Perform the addition: 48 + 93 + 117.
   (You might want to mention that one way to do this calculation is to add the 40 + 90 + 110 = 240 and then to add 8 + 3 + 7 = 18, so that you can get 240 + 18 = 258 as your answer.)

2. **Make Change:** The purchase is $21.08, and the amount tendered is $100.00.
   (You might suggest that your students try making change mentally by adding up to the amount tendered. Then they could check it with pencil-and-paper methods. The answer is 2-P, 1-N, 1-D, 3-Q, 3-$1, 1-$5, 1-$10, and 3-$20.)

Challenge Problem

1. **History of Math:** In ancient Egypt, multiplication was carried out by a process called duplication. For example, to multiply 12 x 23, you would write down the following numbers.

1	23
2	46
4	92
8	184

   Then you would add the numbers in the right column that correspond to the numbers in the left column that add up to 12. Since 4 and 8 add up to 12, you would add 92 + 184 = 276 to get the answer of 276. Try your hand at duplication by multiplying 18 x 21.
   (You should write down the following numbers.

1	21
2	42
4	84
8	168
16	336

   then you would add those numbers in the right column that correspond to the numbers in the left column that add up to 18. Since 2 and 16 add up to 18, you would add 42 + 336 and get 378 as the answer.)

STUDENT PAGE: QUARTER 3: WEEK 7

Student

The theme for week 7 will be making change. You will be given the amount of a purchase and the amount of money given to pay for the purchase. You are to figure the change to be given back by using the least amount of bills and coins in your cash register. The register contains the usual coins (pennies, nickels, dimes, and quarters) and the usual bills (ones, fives, tens, and twenties). All the #2 problems are based on this theme.

Monday   Name: _____

1. **Mental Math:** Calculate 30% of 620.   _____

2. **Make Change:** The purchase is $0.26, and the   _____
   amount tendered is $1.00.

Tuesday   Name: _____

1. **Calculator Math:** Perform the following calculation: $\dfrac{2 - \sqrt{5}}{\sqrt{3}}$.   _____

2. **Make Change:** The purchase is $3.91, and the   _____
   amount tendered is $10.00.

Wednesday   Name: _____

1. **Estimation:** Gary exercises about 40 minutes every day. How   _____
   much does he exercise in a year?

2. **Make Change:** The purchase is $2.19, and the   _____
   amount tendered is $5.04.

99

STUDENT PAGE: QUARTER 3: WEEK 7

Teacher

Thursday   Name: _____

1. **Calculator Math:** Perform the calculation: _____
   $$\frac{(4.75 - 2.93) \times (12.58 + 4.05)}{(7.29 - 3.49)}.$$

2. **Make Change:** The purchase is $6.11, and the   _____
   amount tendered is $11.25.

Friday   Name: _____

1. **Mental Math:** Perform the addition: 48 + 93 + 117.   _____

2. **Make Change:** The purchase is $21.08, and the   _____
   amount tendered is $100.00.

Challenge Problem   Name: _____

1. **History of Math:** In ancient Egypt, multiplication was carried   _____
   out by a process called duplication. For example, to multiply
   12 x 23, you would write down the following numbers,

1	23
2	46
4	92
8	184

   then you would add the numbers in the right column that correspond to the numbers in the left column that add up to 12. Since
   4 and 8 add up to 12, you would add 92 + 184 = 276 to get the
   answer of 276. Try your hand at duplication by multiplying
   18 x 21.

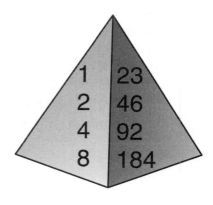

TEACHER PAGE: QUARTER 3: WEEK 8

The theme for week 8 is clock geometry. You will be given a time (assume the hour hand is always right on the number), and you will have to figure out the angle from the big hand to the small hand, plus the distance along the circumference from the big hand to the small hand. We will deal only with the angles between 0 and 180 degrees, and the radius of the clock is 1 unit. All the #2 problems are based on this theme.

Monday

1. **Mental Math:** Multiply: 8 x 37 x 25.
   (You might mention that if there are compatible numbers that they should be multiplied first. In this case either (8 x 25) x 37 or 2 x (4 x 25) x 37 might be the way to look at the problem. Thus we get (2 x 100) x 37 = 2 x 3700 = 7400.)

2. **Clock Geometry:** The time is 3:00. What is the angle, and what is the distance?
   (You might want to review time on a non-digital clock and use the fact that there are 30 degrees between each hour on the clock face. Also, the circumference of this circle of radius 1 is Diameter x π = 2 x π = 6.28…. Therefore, between the small hand and the big hand for 3:00 would be 30 + 30 + 30 = 90 degrees; hence the distance on the circumference would be 90/360 of the 6.28 total circumference, i.e., 1.57.)

Tuesday

1. **Calculator Math:** Perform the following calculation: -2.38 x (-3.11 - 4.59) x (-5.07 - 6.30).
   (You might want to tell your students about the +/- key or the (-) key and the use of parentheses on their calculators. You can perform the following keystrokes: (-) 2.38 x ((-) 3.11 - 4.59) x ((-) 5.07 - 6.30) =. The answer is -208.367.)

2. **Clock Geometry:** The time is 2:35. What is the angle, and what is the distance?
   (You might want to review time on a non-digital clock and use the fact that there are 30 degrees between each hour on the clock face. Also, the circumference of this circle of radius 1 is Diameter x π = 2 x π = 6.28…. Therefore, between the small hand and the big hand for 2:35 would be 30 + 30 + 30 + 30 + 30 = 150 degrees; hence the distance on the circumference would be 150/360 of the 6.28 total circumference, i.e., 2.61799.)

Wednesday

1. **Estimation:** You decide to save $75 a month for three years. Estimate how much you will save.
   (You might want to tell your students to round and then multiply. One way would be to multiply $70 x 40 = $2800. Notice that I rounded the $75 down since I rounded the 36 up to 40.)

2. **Clock Geometry:** The time is 11:45. What is the angle, and what is the distance?
   (You might want to review time on a non-digital clock and use the fact that there are 30 degrees between each hour on the clock face. Also, the circumference of this circle of radius 1 is Diameter x π = 2 x π = 6.28…. Therefore, between the small hand and the big hand for 11:45 would be 30 + 30 = 60 degrees; hence the distance on the circumference would be 60/360 of the 6.28 total circumference, i.e., 1.0472.)

TEACHER PAGE: QUARTER 3: WEEK 8

Thursday

1. **Calculator Math:** You decide to save $75 a month for three years. How much will you save? (You might want to remind your students that there are 12 months in a year and hence 36 months in three years. You can perform the following keystrokes: 3 x 12 x 75 =. The answer is $2700.)

2. **Clock Geometry:** The time is 8:10. What is the angle, and what is the distance? (You might want to review time on a non-digital clock and use the fact that there are 30 degrees between each hour on the clock face. Also, the circumference of this circle of radius 1 is Diameter x π = 2 x π = 6.28…. Therefore, between the small hand and the big hand for 8:10 would be 30 + 30 + 30 + 30 + 30 + 30 = 180 degrees; hence the distance on the circumference would be 180/360 of the 6.28 total circumference, i.e., 3.14159.)

Friday

1. **Mental Math:** Your club is having a breakfast to raise money, and you order 30 dozen eggs. How many eggs is that? (You might need to review the notion of a dozen. Otherwise this is simply 30 x 12 = (3 x 12) x 10 = 36 x 10 = 360.)

2. **Clock Geometry:** The time is 6:30. What is the angle, and what is the distance? (You might want to review time on a non-digital clock and use the fact that there are 30 degrees between each hour on the clock face. Also, the circumference of this circle of radius 1 is Diameter x π = 2 x π = 6.28…. Therefore, between the small hand and the big hand for 6:30 would be 0 degrees; hence the distance on the circumference would be 0/360 of the 6.28 total circumference, i.e., 0.)

Challenge Problems

1. **Clock Geometry:** The time is 3:51. What is the angle, and what is the distance? (You might want to review time on a non-digital clock and use the fact that there are 30 degrees between each hour or 6 degrees between each minute on the clock face. Also, the circumference of this circle of radius 1 is Diameter x π = 2 x π = 6.28…. Therefore, between the small hand and the big hand for 3:51 would be 30 + 30 + 30 + 30 + 24 = 144 degrees; hence the distance on the circumference would be 144/360 of the 6.28 total circumference, i.e., approximately 2.512.)

2. **Clock Geometry:** The time is 11:11. What is the angle, and what is the distance? (You might want to review time on a non-digital clock and use the fact that there are 30 degrees between each hour or 6 degrees between each minute on the clock face. Also, the circumference of this circle of radius 1 is Diameter x π = 2 x π = 6.28…. Therefore, between the small hand and the big hand for 11:11 would be 30 + 30 + 30 + 6 = 96 degrees; hence the distance on the circumference would be 96/360 of the 6.28 total circumference, i.e., approximately 1.6755.)

STUDENT PAGE: QUARTER 3: WEEK 8

Student

The theme for week 8 is clock geometry. You will be given a time (assume the hour hand is always right on the number), and you will have to figure out the angle from the big hand to the small hand, plus the distance along the circumference from the big hand to the small hand. We will deal only with the angles between 0 and 180 degrees, and the radius of the clock is 1 unit. All the #2 problems are based on this theme.

Monday   Name: _____

1. **Mental Math:** Multiply: 8 x 37 x 25.   _____

2. **Clock Geometry:** The time is 3:00. What   _____
   is the angle, and what is the distance?

Tuesday   Name: _____

1. **Calculator Math:** Perform the following calculation: _____
   -2.38 x (-3.11 - 4.59) x (-5.07 - 6.30).

2. **Clock Geometry:** The time is 2:35. What   _____
   is the angle, and what is the distance?

Wednesday   Name: _____

1. **Estimation:** You decide to save $75 a month for three years.   _____
   Estimate how much you will save.

2. **Clock Geometry:** The time is 11:45. What   _____
   is the angle, and what is the distance?

STUDENT PAGE: QUARTER 3: WEEK 8

Student

Thursday   Name: _____

1. **Calculator Math:** You decide to save $75 a month for three years. _____
    How much will you save?

2. **Clock Geometry:** The time is 8:10. What
    is the angle, and what is the distance?   _____

Friday   Name: _____

1. **Mental Math:** Your club is having a breakfast to raise money, _____
    and you order 30 dozen eggs. How many eggs is that?

2. **Clock Geometry:** The time is 6:30. What
    is the angle, and what is the distance?   _____

Challenge Problems   Name: _____

1. **Clock Geometry:** The time is 3:51. What
    is the angle, and what is the distance?   _____

2. **Clock Geometry:** The time is 11:11. What
    is the angle, and what is the distance?   _____

104

TEACHER PAGE: QUARTER 3: WEEK 9

The theme for week 9 is guess that number. All the #2 problems are based on this theme. For each of these problems, you will be given a set of word clues. You will then have to become a mathematical detective to figure out what the number is.

Monday

1. **Mental Math:** Add the following numbers: 230 + 480 + 850.
   (You might want to tell your student to first add the hundreds: 200 + 400 + 800 = 1400; then add the "rest": 30 + 80 + 50 = 160. The answer is 1400 + 160 = 1560.)

2. **Guess That Number:** This number is an abundant whole number (the sum of its proper divisors is greater than the number). The sum of its proper divisors is 36. It is also one less than the square of a small prime number. Guess that number.
   (You might want to review proper divisors and square. You might start with the small prime number 3, and so one less than the square of 3 is 8. But the proper divisors of 8, 1 + 2 + 4 = 6, don't add up to 36. Working your way up the list of small prime numbers, you will eventually find that for 5, one less than its square is 24 and the sum of 24's proper divisors is 1 + 2 + 3 + 4 + 6 + 8 + 12 = 36. The answer is 24.)

Tuesday

1. **Calculator Math:** Perform the following calculation: $(\sqrt{5} + 2)$ x $(\sqrt{5} - 2)$.
   (You might want to tell your students to use the parentheses on their calculators. You can perform the following keystrokes: ($\sqrt{}$ 5 + 2) x ($\sqrt{}$ 5 - 2) =. The answer here is 1.)

2. **Guess That Number:** This number has infinitely many decimal places, and represents the distance around a circle of radius one. Mathematicians have a special symbol for this number. Guess that number.
   (You might want to review the geometry of a circle. The formula for the circumference of a circle is $C = 2$ x r x $\pi = 2$ x 1 x π. The answer is 2π.)

Wednesday

1. **Estimation:** Estimate 47.5% of $2386.
   (You might want to mention that 47.5% is approximately $\frac{1}{2}$, and thus you want $\frac{1}{2}$ of $2400. So your estimate would be $1200.)

2. **Guess That Number:** This number has infinitely many decimal places. It represents the length of the diagonal of a rectangle whose sides are 1 and 2. Guess that number.
   (You might want to review the Pythagorean relationship and the geometry of a rectangle. You get $diagonal^2 = 1^2 + 2^2 = 5$; hence the diagonal is the square root of 5. The answer is $\sqrt{5}$.)

TEACHER PAGE: QUARTER 3: WEEK 9

Thursday

1. **Calculator Math:** Calculate 47.5% of $2386 without using the percent key or the decimal key. (You might want to review multiplication of decimals and see that the decimal points only come in determining where the decimal point is in the final answer. You can perform the following keystrokes: 475 x 2386 = and get 1133350. From the context of the problem, the answer should be in the 1000s. Therefore, the answer is $1133.35.)

2. **Guess That Number:** This is the first perfect whole number (the sum of its proper divisors is equal to the number). It is an even number less than 20. Guess that number.
   (You might want to review what a proper divisor is. Starting with 2, the sum of its proper divisors is 1. Next, the sum of the proper divisors of 3 is 1. Continuing in this manner, you will arrive at 6 and its proper divisors, $1 + 2 + 3 = 6$. The answer is 6.)

Friday

1. **Mental Math:** Perform the following subtraction: $7\frac{3}{7}$ - $2\frac{6}{7}$.

   (You might want to tell your students that if they add the same quantity to both numbers, the answer remains the same. You can add $\frac{1}{7}$ to both numbers to get $7\frac{4}{7}$ - 3 = $4\frac{4}{7}$.)

2. **Guess That Number:** This is a challenging one. The number has infinitely many decimal places, but is the same as one.
   (You might want to show your students that the decimal 0.4999999 repeating is the same as 0.5. To do that, let $n = 0.4999999$. Then $10n = 4.999999$ and $100n = 49.999999$. Hence $100n - 10n = 49.999999 - 4.999999 = 45$. However, $100n - 10n = 90n$, so $90n = 45$ or $n = \frac{45}{90} = \frac{1}{2} = 0.5$. Now a similar thing happens with 0.99999999 repeating and 1.)

Challenge Problems

1. **Guess That Number:** This number has infinitely many decimal places. It is the length of the diagonal of a cube whose sides are of length 1. Guess that number.
   (You might want to draw a cube and look at the diagonal on it. Then notice that this diagonal is the hypotenuse of a right triangle whose legs are 1 and $\sqrt{2}$. Therefore, applying the Pythagorean relationship, we get diagonal$^2 = 1^2 + \sqrt{2}^2 = 1 + 2 = 3$. Therefore, the answer is $\sqrt{3}$.)

2. **Guess That Number:** This number is a fraction whose numerator is 1 and represents the percent 12.5%. Guess that number.
   (You might want to review that % means "divided by 100."

   Thus $12.5\% = \frac{12.5}{100} = \frac{125}{1000} = \frac{1}{8}$.)

106

STUDENT PAGE: QUARTER 3: WEEK 9

Student

The theme for week 9 is guess that number. All the #2 problems are based on this theme. For each of these problems, you will be given a set of word clues. You will then have to become a mathematical detective to figure out what the number is.

Monday   Name: _____

1. **Mental Math:** Add the following numbers: 230 + 480 + 850.   _____

2. **Guess That Number:** This number is an abundant whole number (the sum of its proper divisors is greater than the number). The sum of its proper divisors is 36. It is also one less than the square of a small prime number. Guess that number.   _____

Tuesday   Name: _____

1. **Calculator Math:** Perform the following calculation: _____
    $(\sqrt{5} + 2) \times (\sqrt{5} - 2)$.

2. **Guess That Number:** This number has infinitely many decimal places, and represents the distance around a circle of radius one. Mathematicians have a special symbol for this number. Guess that number.   _____

Wednesday   Name: _____

1. **Estimation:** Estimate 47.5% of $2386.   _____

2. **Guess That Number:** This number has infinitely many decimal places. It represents the length of the diagonal of a rectangle whose sides are 1 and 2. Guess that number.   _____

STUDENT PAGE: QUARTER 3: WEEK 9

Student

Thursday   **Name:** _____

1. **Calculator Math:** Calculate 47.5% of $2386 without using the   _____
    percent key or the decimal key.

2. **Guess That Number:** This is the first perfect      _____
    whole number (the sum of its proper divisors is
    equal to the number). It is an even number less
    than 20. Guess that number.

Friday   **Name:** _____

1. **Mental Math:** Perform the following subtraction: $7\frac{3}{7} - 2\frac{6}{7}$.   _____

2. **Guess That Number:** This is a challenging one.      _____
    The number has infinitely many decimal places,
    but is the same as one.

Challenge Problems   **Name:** _____

1. **Guess That Number:** This number has infinitely many decimal   _____
    places. It is the length of the diagonal of a cube whose sides are
    of length 1. Guess that number.

2. **Guess That Number:** This number is a fraction      _____
    whose numerator is 1 and represents the per-
    cent 12.5%. Guess that number.

108

TEACHER PAGE: QUARTER 4: WEEK 1

The theme for week 1 is travel mathematics. All of the #2 problems are based on this theme.

Monday

1. **Mental Math:** 6 hours for 300 miles is equal to how many miles for 4 hours?
   (You might want to mention that this is simply a proportion problem. You can mentally set up

   $$\frac{4}{6} = \frac{2}{3} = \frac{?}{300}$$; the answer is 200 miles.)

2. **Travel Math:** You are traveling from New York to Washington, D.C., (a distance of 237 miles) at an average speed of 60 miles per hour. How long does it take you to get there?
   (You might want to mention Distance = Rate x Time. In this case you have 237 miles = 60 miles per hour x Time, which means that Time = $\frac{237}{60}$ hours = 3.95 hours or 3 hours 57 minutes.)

Tuesday

1. **Calculator Math:** Perform the following calculation: [48 x (76 - 37) + 92]/(-105 - 53).
   (You might want to talk about the use of parentheses and the +/- key or (-) key on the calculator. You can perform the following keystrokes: (48 x (76 - 37) + 92) / ((-) 105 - 53) =. The answer is -12.43037975.)

2. **Travel Math:** On Interstate 90 it is 159 miles from New York City to Albany, 95 miles from Albany to Utica, 56 miles from Utica to Syracuse, 90 miles from Syracuse to Rochester, and finally 64 miles from Rochester to Buffalo. How far is it from New York City to Buffalo?
   (You might want to talk about road maps and the idea that a journey is the sum of smaller trips. The answer is 159 + 95 + 56 + 90 + 64 = 464 miles.)

Wednesday

1. **Estimation:** Estimate (to the nearest thousand) the following sum: 11,234 + 23,589 + 73,499 + 47,811 + 92,345.
   (You might give students a hint about rounding each number to the nearest thousand and then adding those numbers, so 11,000 + 24,000 + 73,000 + 48,000 + 92,000 = 248,000.)

2. **Travel Math:** It is 598 miles from Amarillo, Texas, to Little Rock, Arkansas, and you make the trip in $10\frac{1}{2}$ hours. What was your average speed for the trip?
   (You might mention Distance = Rate x Time. In this case you have 598 miles = Rate x $10\frac{1}{2}$ hours. Solving for Rate = 598 miles / 10.5 hours, you get approximately 56.95 miles per hour.)

TEACHER PAGE: QUARTER 4: WEEK 1

Thursday

1. **Calculator Math:** Calculate the average age of the following adults: 25, 31, 34, 22, 28, 32, 40, 27.
    (You might mention how to calculate an average. You can perform the following keystrokes: 25 + 31 + 34 + 22 + 28 + 32 + 40 + 27 = / 8 =. The answer is 29.875 = 30.)

2. **Travel Math:** Your parents are driving along on cruise-control at 65 miles per hour for 2 hours 45 minutes before they have to stop for gas. How far did they travel?
    (You might mention Distance = Rate x Time. In this case, Distance = 65 miles per hour x 2.75 hours = 178.75 miles.)

Friday

1. **Mental Math:** Find 25% of $280.
    (You might want to explain how some percents can be thought of more easily in terms of fractions. Thus we have $\frac{1}{4}$ of $280 = $70.)

2. **Travel Math:** For the first 150 miles your parents average 60 miles per hour, but for the next 50 miles they only average 40 miles per hour. What was their average speed for the 200-mile trip?
    (You might mention Distance = Rate x Time. In this problem the students must find the Time for the whole trip by finding the Time for each part of the trip. For the first part, take 150 / 60 = 2.5. For the second part, take 50 / 40 = 1.25. So, Time for the whole trip is equal to 3.75 hours. Therefore, the Rate = 200 miles / 3.75 hours = 53.$\overline{3}$ miles per hour.)

Challenge Problem

1. **Travel Math:** You are flying 1,100 miles to see your cousins in California. If the plane averages 300 miles an hour, what is the longest your layover in Denver can last if you plan to get to Seattle in 5 hours?
    (You will want to use Distance = Rate x Time to figure that the flying time of the trip is Distance divided by Rate or $\frac{1100}{600} = 3\frac{2}{3} = 3$ hours 40 minutes. Subtract that time from 5 hours to determine that you will have 1 hour 20 minutes for the layover.)

STUDENT PAGE: QUARTER 4: WEEK 1

Student

The theme for week 1 is travel mathematics. All of the #2 problems are based on this theme.

Monday   Name: _____

1. **Mental Math:** 6 hours for 300 miles is equal to how many miles   _____
    for 4 hours?

2. **Travel Math:** You are traveling from New York to Washington,   _____
    D.C., (a distance of 237 miles) at
    an average speed of 60 miles per
    hour. How long does it take you to
    get there?

Tuesday   Name: _____

1. **Calculator Math:** Perform the following calculation: _____
    [48 x (76 - 37) + 92]/(-105 - 53).

2. **Travel Math:** On Interstate 90 it is 159 miles from New York City   _____
    to Albany, 95 miles from Albany to Utica, 56 miles from Utica to
    Syracuse, 90 miles from Syracuse to Rochester, and finally 64
    miles from Rochester to Buffalo.
    How far is it from New York City to
    Buffalo?

Wednesday   Name: _____

1. **Estimation:** Estimate (to the nearest thousand) the following   _____
    sum: 11,234 + 23,589 + 73,499 + 47,811 + 92,345.

2. **Travel Math:** It is 598 miles from Amarillo, Texas, to Little Rock,   _____
    Arkansas, and you make the trip in
    $10\frac{1}{2}$ hours. What was your average
    speed for the trip?

STUDENT PAGE: QUARTER 4: WEEK 1

Student

Thursday   Name: _____

1. **Calculator Math:** Calculate the average age of the following _____
    adults: 25, 31, 34, 22, 28, 32, 40, 27.

2. **Travel Math:** Your parents are driving along on cruise-control at _____
    65 miles per hour for 2 hours 45
    minutes before they have to stop
    for gas. How far did they travel?

Friday   Name: _____

1. **Mental Math:** Find 25% of $280.   _____

2. **Travel Math:** For the first 150 miles your parents average 60 _____
    miles per hour, but for the next 50 miles they only average 40
    miles per hour. What was their av-
    erage speed for the 200-mile trip?

Challenge Problem   Name: _____

1. **Travel Math:** You are flying 1,100 miles to see your cousins in _____
    California. If the plane averages 300 miles an hour, what is the
    longest your layover in Denver can last if you plan to get to Se-
    attle in 5 hours?

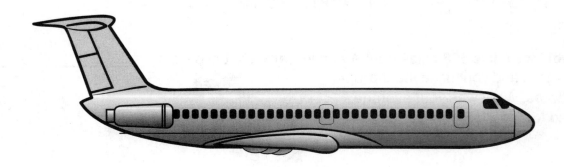

TEACHER PAGE: QUARTER 4: WEEK 2

The theme of week 2 will be soccer geometry. All of the #2 problems will be based on this theme.

Monday

1. **Mental Math:** Multiply the following numbers: 34 x 98.
   (You might want to mention that one way of thinking about this problem is to think of the answer as one hundred 34s minus two 34s. This would make the answer 3400 - 68 = 3332.)

2. **Soccer Geometry:** A soccer field is rectangular in shape. The adult-size soccer field must be at least 90 meters, but no more than 120 meters, long and at least 45 meters, but no more than 90 meters, wide. What is the area of the smallest possible soccer field and the largest possible soccer field?
   (You might want to review the area of a rectangle. The smallest field has an area of 90 x 45 = 4,050 square meters, and the largest field has an area of 120 x 90 = 10,800 square meters.)

Tuesday

1. **Calculator Math:** A meal combo at the local fast food restaurant costs $4.29. If sales tax is 7.75%, then what is your total bill for the combo meal?
   (You might want to review percent and then mention that you can figure the total bill by adding the tax onto the price or by figuring 107.75% of the $4.29. You can perform the following keystrokes: 4.29 + 4.29 x .0775 = or 4.29 x 1.0775 =. In either case, the answer is $4.62.)

2. **Soccer Geometry:** There is a circle at the center of the field. It is to be of radius 9.15 meters. What is its area and circumference?
   (You might want to review the area and circumference formulas for a circle. The value of π (pi) is 3.14159…. You get *area* = r^2 x π = $(9.15)^2$ x π = 263.0 square meters and *circumference* = 2 x r x π = 2 x 9.15 x π = 57.5 meters.)

Wednesday

1. **Estimation:** Estimate the quotient in the following division problem: 6279 / 7.15.
   (You might want to suggest rounding each number to an "appropriate" number. In this problem, you could estimate with 6300 / 7 = 900.)

2. **Soccer Math:** The face of each goal is a rectangle that is 7.32 meters wide and 2.44 meters high. The goalie must defend what area in square feet? (Remember, 1 meter = 3.28 feet.)
   (You might want to tell your students to convert the dimensions to feet and then calculate the area of the rectangular face. The width is 7.32 x 3.28 = 24.0096 feet, and the length is 2.44 x 3.28 = 8.0032 feet; hence, the area is 24.0096 x 8.0032 = 192.154 square feet.)

TEACHER PAGE: QUARTER 4: WEEK 2

Teacher

Thursday

1. **Calculator Math:** Find the quotient of the following division problem without using the decimal point on your calculator: 6279 / 7.15. (Round to the nearest thousandths place.)
    (You might want to review how the long division is done by hand as a hint for this problem. You can perform the following keystrokes: 6279 / 715 = and get 8.78182. But since the divisor has two decimal places, we need to move the decimal point over two places to the right. Thus the answer is 878.182.)

2. **Soccer Geometry:** The penalty area is a rectangle at each end of the field in front of the goal. The length of the rectangle is 18 + 8 + 18 yards and the width is 18 yards. What is the area of the penalty area in square feet?
    (You might want to review the area of a rectangle and the fact that there are 3 feet in 1 yard. You get the length = (18 + 8 + 18) x 3 = 132 feet and the width = 18 x 3 = 54 feet; hence, the area is 132 x 54 = 7,128 square feet.)

Friday

1. **Mental Math:** Perform the following subtraction: 618 - 93.
    (You might mention that one way of thinking of this problem is to think 618 - 100 = 518, but I took away 7 too many. Thus the answer is 518 + 7 = 525.)

2. **Soccer Geometry:** At each corner of the soccer field is a quarter circle of radius 1 meter. Draw a picture of what one of these corner arcs would look like and find its area.
    (You might want to review the area of a circle and talk about what part of the whole the corner arc represents. The area is $\frac{1}{4}$ of the whole area of the circle, which is $\frac{1}{4}$ x r^2 x $\pi = \frac{1}{4}$ x 1 x π = 0.785 square meters.)

Challenge Problem

1. **Geometry:** What is the area of the shaded region in the picture below? The square is two units long on each side.
    (You might want to review the area of a circle and square, and discuss how some areas can be figured by "whole" area minus "hole" area. In this case the "whole" area is the square and the "hole" is the two halves of the circle. The answer is 4 - 3.14159 = 0.85841.)

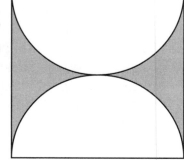

STUDENT PAGE: QUARTER 4: WEEK 2

The theme of week 2 will be soccer geometry. All of the #2 problems will be based on this theme.

Monday   Name: _____

1. **Mental Math:** Multiply the following numbers: 34 x 98.   _____

2. **Soccer Geometry:** A soccer field is rectan-   _____
    gular in shape. The adult-size soccer field
    must be at least 90 meters, but no more than
    120 meters, long and at least 45 meters, but
    no more than 90 meters, wide. What is the
    area of the smallest possible soccer field and
    the largest possible soccer field?

Tuesday   Name: _____

1. **Calculator Math:** A meal combo at the local fast food restaurant   _____
    costs $4.29. If sales tax is 7.75%, then what is your total bill for
    the combo meal?

2. **Soccer Geometry:** There is a circle at the   _____
    center of the field. It is to be of radius 9.15
    meters. What is its area and circumference?

Wednesday   Name: _____

1. **Estimation:** Estimate the quotient in the following division prob-   _____
    lem: 6279 / 7.15.

2. **Soccer Math:** The face of each goal is a rect-   _____
    angle that is 7.32 meters wide and 2.44
    meters high. The goalie must defend what
    area in square feet? (Remember, 1 meter =
    3.28 feet.)

STUDENT PAGE: QUARTER 4: WEEK 2

Student

Thursday   Name: _____

1. **Calculator Math:** Find the quotient of the following division prob-   _____
   lem without using the decimal point on your calculator:
   6279 / 7.15. (Round to the nearest thousandths place.)

2. **Soccer Geometry:** The penalty area is a rect-      _____
   angle at each end of the field in front of the
   goal. The length of the rectangle is 18 + 8 +
   18 yards and the width is 18 yards. What is
   the area of the penalty area in square feet?

Friday   Name: _____

1. **Mental Math:** Perform the following subtraction: 618 - 93.   _____

2. **Soccer Geometry:** At each corner of the soc-      _____
   cer field is a quarter circle of radius 1 meter.
   Draw a picture of what one of these corner
   arcs would look like and find its area.

Challenge Problem   Name: _____

1. **Geometry:** What is the area of the shaded region in the picture   _____
   below? The square is two units long on each side.

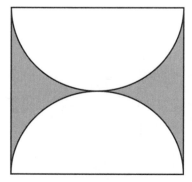

TEACHER PAGE: QUARTER 4: WEEK 3

The theme for week 3 is American Literature. All of the #2 problems will be based on this theme.

Monday

1. **Mental Math:** Perform the following division: 4800 / 16.
   (You might want to tell your students that if you multiply each number in a division problem by the same number, then the answer remains the same, and if you multiply the numerator by a number, then the answer changes by a factor of that number. In this problem, you can mentally do $\frac{48}{16} = 3$ and then $\frac{480}{16} = 30$ and finally, $\frac{4800}{16} = 300$.)

2. **American Literature:** Samuel Clemens (Mark Twain) was the author of the novel *The Adventures of Huckleberry Finn.* This great American novel, written about nineteenth-century life on and around the Mississippi River, was first published in the year that has 13 and 29 as factors. What is the year?
   (You might want to review what it means for a number to be a factor of another number. Thus the year = 13 x 29 x ? = 377 x ?. With a little guess-and-check, you see that year = 377 x 5 = 1885 must be the correct answer.)

Tuesday

1. **Calculator Math:** Perform the following calculation: $[(3.47 \times 10^4) \times (6.02 \times 10^6)] / (5.32 \times 10^7)$.
   (You might want to show your students how to use the scientific notation feature on their calculators. You can perform the following keystrokes: 3.47 EE 4 x 6.02 EE 6 = / 5.32 EE 7 =. The answer is 3926.58.)

2. **American Literature:** *The Grapes of Wrath* was a novel written by John Steinbeck. This American novel depicts life in the Dust Bowl days in America's southwest. It was first published in the year three more than forty-four squared. What is the year?
   (You might want to talk about how to translate a word sentence into numbers and number operations. Thus you get that the year = 3 + (44 x 44) = 1939.)

Wednesday

1. **Estimation:** Estimate the perimeter of a square whose area is 441.
   (You might want to review the area and perimeter of a square. A crude way of estimating square roots is to think 10 x 10 = 100, 20 x 20 = 400, 30 x 30 = 900, and so on. In this case, the side is approximately 20, and hence the perimeter is 80.)

2. **American Literature:** Alice Walker was the author of the novel *The Color Purple.* This novel was the basis for the movie by the same name that starred Whoopi Goldberg. The book was first published in the year whose square is 71,676 less than 4 million. What is the year?
   (You might want to talk about how to translate words into numbers and number operations. You might also want to review the relationship between square and square root. Thus you have year^2 = 4,000,000 - 71,676 = 3,928,324, and taking the square root you get year = 1982.)

           117

TEACHER PAGE: QUARTER 4: WEEK 3

Thursday

1. **Calculator Math:** Calculate the perimeter of a square whose area is 441.
   (You might want to review the area and perimeter of a square. Since area equals side squared, then the side is the square root of 441 or 21. Thus the perimeter is 21 + 21 + 21 + 21 = 84.)

2. **American Literature:** The novel *Little Women* was written by Louisa May Alcott. This nineteenth-century work was first published in the year whose sum of digits is 23 and whose units digit is the same as its hundreds digit. What is the year?
   (You might want to talk about problem-solving strategies and setting up equations. You get the equations $th + h + te + u = 23$ and $h = u$. Using the clue that the book was written in the nineteenth century, you know that $th = 1$ and $h = 8$. Thus you can easily deduce $u = 8$ and $te = 6$. The answer is 1868.)

Friday

1. **Mental Math:** Perform the following multiplication: 47.82 x 10,000.
   (You might want to review that multiplication by a power of ten simply means moving the decimal point an appropriate number of places. In the case, move the decimal point four places to the right, so the answer is 478,200.)

2. **American Literature:** *The Catcher in the Rye* was written by J. D. Salinger. This book about three days in the life of a troubled boy caused quite a stir when it was published. In fact, it has been banned in some school libraries. The book was first published in the first prime-numbered year of the second half of the twentieth century. What is the year?
   (You might want to tell students where they should start looking for the prime number. It won't be hard, since the year is 1951.)

Challenge Problems

1. **American Literature:** *The Scarlet Letter* is the Nathaniel Hawthorne novel written about the values and morals of life in colonial New England. This novel was first published in the year that has prime factors of 2, 5, and 37. What is the year?
   (You need to review the concept of prime factor. You then know that 37 x 5 x 2 = 370 is a factor of the number. But that is obviously not the year. So there must be another factor of 2, 5, or 37 in the number. Through trial-and-error you can figure that there must be another factor of 5 in the number. So the year is 1850.)

2. **American Literature:** Twentieth-century author Margaret Mitchell won the Pulitzer Prize for *Gone With the Wind,* the novel that inspired the movie that won the Academy Award for best picture in 1939. This novel was published in the year whose sum of digits is 19 and the tens digit is one-third the hundreds digit. What is the year?
   (You can tell from the clue that the novel was written in the 1900s. From there you can get the tens digit is 3. By setting up the equation $1 + 9 + 3 + u = 19$, you can determine that the units digit is 6. The year is 1936.)

STUDENT PAGE: QUARTER 4: WEEK 3    **Student**

The theme for week 3 is American Literature. All of the #2 problems will be based on this theme.

Monday   **Name:** _____

1. **Mental Math:** Perform the following division: 4800 / 16.   _____

2. **American Literature:** Samuel Clemens (Mark Twain) was the   _____
   author of the novel *The Adventures of Huckleberry Finn.* This
   great American novel, written about nine-
   teenth-century life on and around the Missis-
   sippi River, was first published in the year that
   has 13 and 29 as factors. What is the year?

Tuesday   **Name:** _____

1. **Calculator Math:** Perform the following calculation: _____
   $[(3.47 \times 10^4) \times (6.02 \times 10^6)] / (5.32 \times 10^7)$.

2. **American Literature:** *The Grapes of Wrath* was a novel written   _____
   by John Steinbeck. This American novel depicts
   life in the Dust Bowl days in America's southwest. It was first published in the year three
   more than forty-four squared. What is the year?

Wednesday   **Name:** _____

1. **Estimation:** Estimate the perimeter of a square whose area is   _____
   441.

2. **American Literature:** Alice Walker was the author of the novel   _____
   The Color Purple. This novel was the basis for the movie by the
   same name that starred Whoopi
   Goldberg. The book was first published
   in the year whose square is 71,676 less
   than 4 million. What is the year?

119

STUDENT PAGE: QUARTER 4: WEEK 3

Thursday   Name: _____

1. **Calculator Math:** Calculate the perimeter of a square whose _____
   area is 441.

2. **American Literature:** The novel *Little Women* was written by _____
   Louisa May Alcott. This nineteenth-century work
   was first published in the year whose sum of digits is 23 and whose units digit is the same as its
   hundreds digit. What is the year?

Friday   Name: _____

1. **Mental Math:** Perform the following multiplication: _____
   47.82 x 10,000.

2. **American Literature:** *The Catcher in the Rye* was written by _____
   J. D. Salinger. This book about three days in the life of a troubled
   boy caused quite a stir when it was published.
   In fact, it has been banned in some school libraries. The book was first published in the first
   prime-numbered year of the second half of the
   twentieth century. What is the year?

Challenge Problems   Name: _____

1. **American Literature:** *The Scarlet Letter* is the Nathaniel _____
   Hawthorne novel written about the values and
   morals of life in colonial New England. This
   novel was first published in the year that has
   prime factors of 2, 5, and 37. What is the year?

2. **American Literature:** Twentieth-century author Margaret _____
   Mitchell won the Pulitzer Prize for *Gone With the Wind,* the novel
   that inspired the movie that won the Academy Award for best
   picture in 1939. This novel was published in the year whose
   sum of digits is 19 and the tens digit is one-third the hundreds
   digit. What is the year?

120

TEACHER PAGE: QUARTER 4: WEEK 4

The theme for week 4 is American Presidents. All of the #2 problems are based on this theme.

Monday

1. **Mental Math:** Multiply 30 x 140.
   (You might want to tell your students that you can think of the above problem as (3 x 14) x 10 x 10 = 42 x 10 x 10 = 4,200.)

2. **American Presidents:** John Adams was the second President of the United States. His term of office began in the year whose sum of digits is 24 and whose units and hundreds digit are both the same, two less than the tens digit. In what year did his term begin?
   (You might want to review problem-solving strategies and setting up equations. You get the equations $th + h + te + u = 24$, $h = u$, and $te - 2 = h$. You should know that the year was either in the late 1700s or the early 1800s; hence $th = 1$ and $h = 7$ or 8. With a little guess-and-check, you can deduce that $h = 7$, $te = 9$, and $u = 7$. The answer is 1797.)

Tuesday

1. **Calculator Math:** Add the following list of numbers without using the decimal point key on your calculator: 15.78 + 8.6 + 117 + 43.9.
   (You might want to review how you would do that by hand and then how you could do it with the calculator without the decimal point key. So, if you add 1578 + 860 + 11700 + 4390 to get 18528, then your answer is 185.28.)

2. **American Presidents:** Grover Cleveland was the 22nd and the 24th President of the United States. His first term in office began in the year whose square is 3,553,225. In what year did Cleveland's first term begin?
   (You might want to review the relationship between square and square root. Thus year^2 = 3,553,225, and by taking the square root we get year = 1885.)

Wednesday

1. **Estimation:** Your watch gains 17 seconds each day. How much time will it gain in a year?
   (You might want to mention that this estimation can probably best be done in terms of minutes. You gain approximately $\frac{1}{4}$ of a minute each day for approximately 360 days. Thus your watch gains about 90 minutes in a year.)

2. **American Presidents:** Theodore Roosevelt was president from 1901 to 1909. He was the *n*th numbered president, where *n* is found by finding the square of five and then adding one.
   (You might want to review translating from words to numbers and number operations. You get $n = 5 \times 5 + 1 = 26$. Thus Teddy Roosevelt was the 26th President of the United States.)

TEACHER PAGE: QUARTER 4: WEEK 4

Thursday

1. **Calculator Math:** Your watch gains 17 seconds each day. How much time will it gain in a year? (You might want to instruct your students to calculate the number of seconds gained in a year and then convert to hours and minutes. The answer is 17 x 365 = 6,205 seconds or 103.4 minutes or 1.72 hours.)

2. **American Presidents:** Herbert Hoover was the 31st President of the United States. He served only one term, which occurred during the Great Depression. Hoover took office in the year whose sum of digits is 21 and whose units and hundreds digits are the same. In what year did Hoover become president?
(You might want to review problem-solving strategies and setting up equations. You get the equations $th + h + te + u = 21$ and $h = u$. You should know that the year was in the 1900s. If $th = 1$ and $h = 9$, then $u = 9$ and $te = 2$. The answer is 1929.)

Friday

1. **Mental Math:** If candy bars are on sale for 2 for $0.96, then how many could you buy with $4.80?
(You might want to mention that this can be done by looking at the price per candy bar ($0.48) and then seeing that there are exactly 10 of those in $4.80. The answer is 10 candy bars.)

2. **American Presidents:** Franklin D. Roosevelt was the 32nd President of the United States. He served the most terms of any president in history. He died during his fourth term. His death was in the year whose units digit and tens digit add up to nine, and the units digit is one more than the tens digit. In what year did Roosevelt die?
(You might want to review problem-solving strategies and setting up equations. You get the equations $te + u = 9$ and $u = te + 1$. With a little guess-and-check, you can figure $te = 4$ and $u = 5$, so the year must be 1945.)

Challenge Problem

1. **American Presidents:** Andrew Jackson was most famous as a general in the War of 1812 where he helped the United States defeat the British in the battle of New Orleans. He became President of the United States from the state of Tennessee in the year that can be found by cubing twelve and adding one hundred one. What is the year?
(This is an exercise in decoding words into numbers and number operations. You might want to review what a cube is. The answer is (12 x 12 x 12) + 101 = 1728 + 101 = 1829.)

STUDENT PAGE: QUARTER 4: WEEK 4

Student

The theme for week 4 is American Presidents. All of the #2 problems are based on this theme.

Monday   Name: _____

1. **Mental Math:** Multiply 30 x 140.   _____

2. **American Presidents:** John Adams' term of
   office began in the year whose sum of digits is   _____
   24 and whose units and hundreds digit are both
   the same, two less than the tens digit. In what
   year did his term begin?

J. ADAMS

Tuesday   Name: _____

1. **Calculator Math:** Add the following list of numbers without us-   _____
   ing the decimal point key on your calculator:
      15.78 + 8.6 + 117 + 43.9.

2. **American Presidents:** Grover Cleveland was
   the 22nd and the 24th President of the United   _____
   States. His first term in office began in the year
   whose square is 3,553,225. In what year did
   Cleveland's first term begin?

G. CLEVELAND

Wednesday   Name: _____

1. **Estimation:** Your watch gains 17 seconds each day. How much   _____
   time will it gain in a year?

2. **American Presidents:** Theodore Roosevelt
   was president from 1901 to 1909. He was the   _____
   nth numbered president, where n is found by
   finding the square of five and then adding one.

T. ROOSEVELT

123

STUDENT PAGE: QUARTER 4: WEEK 4   **Student**

Thursday   Name: _____

1. **Calculator Math:** Your watch gains 17 seconds each day. How   _____
    much time will it gain in a year?

2. **American Presidents:** Herbert Hoover served
    only one term—during the Great Depression.
    Hoover took office in the year whose sum of
    digits is 21 and whose units and hundreds dig-
    its are the same. In what year did Hoover be-
    come president?

H. HOOVER

Friday   Name: _____

1. **Mental Math:** If candy bars are on sale for 2 for $0.96, how   _____
    many could you buy with $4.80?

2. **American Presidents:** Franklin D. Roosevelt
    served the most terms of any president in his-
    tory. He died during his fourth term. His death
    was in the year whose units digit and tens digit
    add up to nine, and the units digit is one more
    than the tens digit. In what year did Roosevelt
    die?

F. ROOSEVELT

Challenge Problem   Name: _____

1. **American Presidents:** Andrew Jackson was
    most famous as a general in the War of 1812
    where he helped the United States defeat the
    British in the battle of New Orleans. He became
    President of the United States from the state
    of Tennessee in the year that can be found by
    cubing twelve and adding one hundred one.
    What is the year?

A. JACKSON

TEACHER PAGE: QUARTER 4: WEEK 5

The theme for week 5 is famous women mathematicians. All of the #2 problems are based on this theme.

Monday

1. **Mental Math:** Subtract: 12.46 - 8.98.
   (You might want to mention that you can add the same quantity to both numbers and keep the answer the same in this calculation; so if you add 0.02 to each number, you get 12.48 - 9 = 3.48.)

2. **Women Mathematicians:** Sophie Germain was a French woman born in 1776. She was not allowed to study mathematics formally because she was a woman. Sophie had to study on her own, but she became one of the great mathematicians of her time. She died in the year whose sum of digits is 13, and its tens digit is triple its units digit. In what year did she die?
   (You might review problem-solving strategies and setting up equations. You get the equations $th + h + te + u = 13$ and $te = 3 \times u$. Guessing that she died in the 1800s, you have $th = 1$ and $h = 8$; so the only possibility for the other two digits is for $te = 3$ and $u = 1$. Sophie died in 1831.)

Tuesday

1. **Calculator Math:** Find the area of the surface of the earth. The radius of the earth is approximately 3,950 miles.
   (You might want to mention the formula for the surface area of a sphere and how to use the π key on the calculator. The surface area = $4 \times r \times r \times \pi$. You can perform the following keystrokes: $4 \times 3950 \times 3950 \times \pi$. The answer is 196,066,797.5 square miles.)

2. **Women Mathematicians:** Emmy Noether was a German woman born in 1882. She had trouble earning her rightful place in the academic world because of her gender. She was forced to leave Germany when Hitler came into power. When you add 3 to this year its square root is 44. In what year did she leave Germany?
   (You might want to talk about setting up an equation in this situation. It would look like $\sqrt{\text{year} + 3} = 44$; thus year is the square of 44 less 3. The year is 1933.)

Wednesday

1. **Estimation:** Find a high and low estimate for the following product: 381 x 625.
   (You might want to discuss rounding both of the numbers down to get a low estimate and rounding them both up to get a high estimate. In the above problem, that would be 300 x 600 = 180,000 for the low estimate and 400 x 700 = 280,000 for the high estimate.)

2. **Women Mathematicians:** Grace Chisholm Young was an Englishwoman born in 1868. She became the first woman granted a doctorate in Germany. She returned to England, where she and her husband became famous mathematicians. She died in the year that has five factors of 3 and three factors of 2. In what year did she die?
   (You might want to talk about what it means to be a factor. Thus the year = 3 x 3 x 3 x 3 x 3 x 2 x 2 x 2 x ? = 1944 x ?. But only ? = 1 makes sense, so the year would be 1944.)

TEACHER PAGE: QUARTER 4: WEEK 5   Teacher

Thursday

1. **Calculator Math:** Calculate the following sum: 23.4 + 85.7 + 91.7 + 72.8 + 101.3 + 64.9 + 9.9. (You might have all the students start at a common signal and give them a time limit. Also, you might want to tell them the appropriate way to add a list of numbers with their calculators. You can perform the following keystrokes: 23.4 + 85.7 + 91.7 + 72.8 + 101.3 + 64.9 + 9.9 =. The answer is 449.7)

2. **Women Mathematicians:** Evelyn Boyd Granville was the first African-American woman to earn a Ph.D. in mathematics. She received this degree from Yale University in 1949. Evelyn Granville was born 24 years past the year whose square is 3,610,000. In what year was she born?
   (You might want to mention setting up an equation for this problem. This equation would be Year = $24 + \sqrt{3,610,000} = 1924$.)

Friday

1. **Mental Math:** Multiply 8 x 20.5.
   (You might want to mention how the distributive property can be used to make this a mental computation, i.e., 8 x (20 + 0.5) = 8 x 20 + 8 x 0.5 = 160 + 4 = 164.)

2. **Women Mathematicians:** Sun-Yung Alice Chang is a Chinese-American woman who earned a Ph.D. in mathematics from the University of California, Berkeley in 1974. In 1995 she won a prize for outstanding research in mathematics by a woman. She was born in the year whose sum of digits is 22 and where the units digit is twice the tens digit. In what year was she born? (You might want to review problem-solving strategies and setting up equations. You get the equations $th + h + te + u = 22$ and $u = 2 \times te$. You can assume that the year was in the 1900s, so $th = 1$ and $h = 9$. Thus $te + u = 12$ and $u = 2 \times te$. Using guess-and-check, you get $te = 4$ and $u = 8$. The answer is 1948.)

Challenge Problem

1. **History of Math:** In ancient Mesopotamia (today this is the country of Iraq), a base 60 positional number system was developed. It was very advanced for its time. In a base 60 positional number system, the place values are units, sixties, sixties-squared, and so on, instead of units, tens, hundreds (tens-squared), and so on as in our base 10 positional system. For example, the base 60 number 1, 20, 3 would represent 3 ones plus 20 sixties plus 1 sixty-squared, which would be 3 x 1 + 20 x 60 + 1 x 3600 = 3 + 1200 + 3600 = 4803. Convert our base 10 number 3723 into a base 60 number.
   (You have to begin by figuring out which is the highest place value that goes into the number. Here it would be 3600, which goes into 3723 one time with 123 left over. So the sixty-squared digit has been determined to be 1. Now, how many of the next-smaller place value goes into the left over 123? The answer would be that 60 goes into it 2 times with 3 left over. Thus the sixties digit is 2. Since there are only 3 left over, the units digit will be 3. The answer is 1, 2, 3.)

126

STUDENT PAGE: QUARTER 4: WEEK 5   **Student**

The theme for week 5 is famous women mathematicians. All of the #2 problems are based on this theme.

Monday   Name: _____

1. **Mental Math:** Subtract: 12.46 - 8.98.   _____

2. **Women Mathematicians:** Sophie Germain was a French woman   _____
   born in 1776. She was not allowed to study mathemat-
   ics formally because she was a woman. Sophie
   had to study on her own, but she became one of
   the great mathematicians of her time. She died in
   the year whose sum of digits is 13, and its tens digit
   is triple its units digit. In what year did she die?

Tuesday   Name: _____

1. **Calculator Math:** Find the area of the surface of the earth. The   _____
   radius of the earth is approximately 3,950 miles.

2. **Women Mathematicians:** Emmy Noether was a German woman   _____
   born in 1882. She had trouble earning her rightful place
   in the academic world because of her gender. She
   was forced to leave Germany when Hitler came into
   power. When you add 3 to this year its square root is
   44. In what year did she leave Germany?

Wednesday   Name: _____

1. **Estimation:** Find a high and low estimate for the following prod-   _____
   uct: 381 x 625.

2. **Women Mathematicians:** Grace Chisholm Young was an   _____
   Englishwoman born in 1868. She became the first
   woman granted a doctorate in Germany. She returned
   to England, where she and her husband became famous mathematicians. She died in the year that has
   five factors of 3 and three factors of 2. In what year did
   she die?

STUDENT PAGE: QUARTER 4: WEEK 5

Student

Thursday   Name: _____

1. **Calculator Math:** Calculate the following sum:
   23.4 + 85.7 + 91.7 + 72.8 + 101.3 + 64.9 + 9.9.

2. **Women Mathematicians:** Evelyn Boyd Granville was the first African-American woman to earn a Ph.D. in mathematics. She received this degree from Yale University in 1949. Evelyn Granville was born 24 years past the year whose square is 3,610,000. In what year was she born?

Friday   Name: _____

1. **Mental Math:** Multiply 8 x 20.5.

2. **Women Mathematicians:** Sun-Yung Alice Chang is a Chinese-American woman who earned a Ph.D. in mathematics from the University of California, Berkeley in 1974. In 1995 she won a prize for outstanding research in mathematics by a woman. She was born in the year whose sum of digits is 22 and where the units digit is twice the tens digit. In what year was she born?

Challenge Problem   Name: _____

1. **History of Math:** In ancient Mesopotamia (today this is the country of Iraq), a base 60 positional number system was developed. It was very advanced for its time. In a base 60 positional number system, the place values are units, sixties, sixties-squared, and so on, instead of units, tens, hundreds (tens-squared), and so on as in our base 10 positional system. For example, the base 60 number 1, 20, 3 would represent 3 ones plus 20 sixties plus 1 sixty-squared, which would be 3 x 1 + 20 x 60 + 1 x 3600 = 3 + 1200 + 3600 = 4803. Convert our base 10 number 3723 into a base 60 number.

TEACHER PAGE: QUARTER 4: WEEK 6

The theme for week 6 is famous national parks and their zip codes. We will think of a zip code as being composed of a two-digit number, followed by a one-digit number, followed by another two-digit number, e.g., 63435 would be 63, 4, and 35. All of the #2 problems are based on this theme.

Monday

1. **Mental Math:** Perform the following multiplication: 195.38 x 10,000.
(You might want to remind your students that multiplying by a power of ten means simply moving the decimal place an appropriate number of places. In this problem, moving the decimal place four places to the right results in an answer of 1,953,800.)

2. **National Parks:** Carlsbad Caverns is in New Mexico. It contains the deepest limestone cave in America, going down to a depth of 1,567 feet. The sum of the first and third numbers in the zip code is 108, and their difference, first minus third, is 68. The second number is the first prime number. What is the zip code?
(You might want to review problem-solving strategies and setting up equations. Remember! One is not a prime number. You get the equations $f + t = 108$ and $f - t = 68$. With a little guess-and-check, you can find $f = 88$ and $t = 20$, and the first prime is 2. The answer is 88220.)

Tuesday

1. **Calculator Math:** Calculate the following: $\left(1 + \dfrac{0.06}{365}\right)^{5(365)}$

(You might want to show your students how to use the exponential function on their calculators. You can do this calculation easily by doing the inside first and then the exponential last. You can perform the following keystrokes: .06 / 365 = + 1 = ^ (5 x 365) =. The answer is 1.34983.)

2. **National Parks:** In Alaska, Denali National Park is home to Mount McKinley, the highest peak in North America. The first number in the zip code has factors of 9 and 11, and the third number has factors of 5 and 11. The second number is the average of 5 and 9. What is the zip code?
(You might want to review the concepts of factor and average. The first number = 9 x 11 = 99, the second number is clearly 7, and the third number = 5 x 11 = 55. The answer is 99755.)

Wednesday

1. **Estimation:** Estimate the following: 953 / 18.
(You might want to tell your students to round both numbers to compatible numbers. In this case, 960 / 20 = 48.)

2. **National Parks:** Volcanoes National Park in Hawaii spans some 217,000 acres going from sea level to an elevation of 13,677 feet at the summit of Mauna Loa. This park contains Kilauea, the world's most active volcano. The first number in the zip code has five factors of 2 and one factor of 3. The third number has two factors of 3 and one factor of 2. The second number is the biggest prime less than 10. What is the zip code?
(You might want to review factors and factorizations. The first number = 2 x 2 x 2 x 2 x 2 x 3 = 96, the second number is 7, and the third number = 3 x 3 x 2 = 18. The answer is 96718.)

TEACHER PAGE: QUARTER 4: WEEK 6   

Thursday

1. **Calculator Math:** Perform the following addition: $129.75 + $580.54 + $329.40 + $814.28 + $492.46.
   (You might want to start everyone on a signal and give them a time limit to do this calculation. You can perform the following keystrokes: 129.75 + 580.54 + 329.40 + 814.28 + 492.46 =. The answer is $2346.43.)

2. **National Parks:** Western Colorado is home to the Dinosaurs National Monument. In 1909 Earl Douglass discovered the fossils of dinosaurs at this site. Today there are over 1,500 fossil bones in this park. The first number in the zip code is the square of nine, and the third number is one more than nine. The second number is the first perfect number. What is the zip code?
   (You might want to review translating words into numbers and number operations. Also, you will have to talk about perfect numbers. They are whole numbers whose sum of proper divisors equals the number itself. The first number = 9 x 9 = 81, the third number = 9 + 1 = 10, and the second number is 6 (the sum of proper divisors 1 + 2 + 3 = 6). The answer here is 81610.)

Friday

1. **Mental Math:** Perform the following addition: 15.78 + 21.97.
   (You might mention that if you add something to one number and subtract it from the other number, then the answer remains the same. In this problem, you might add 0.03 to the second number and subtract it from the first, resulting in 15.75 + 22 = 37.75.)

2. **National Parks:** Big Bend National Park is 801,000 acres in a remote part of west Texas where the Rio Grande River makes a "big turn." The nearest town is over 100 miles away. The first number in the zip code is the biggest prime less than 80. The second is the cube of two, and the third number is an even multiple of 17 that is less than 50. What is the zip code?
   (You might want to review the concept of prime numbers, multiples of a number, and the concept of cube. Starting with 79, you find that it is a prime, so the first number is 79. The second number = 2 x 2 x 2 = 8, and the third number = 2 x 17 = 34. The answer is 79834.)

Challenge Problem

1. **National Parks:** Black Canyon of the Gunnison National Park is located in southwestern Colorado along the Gunnison River. The park covers some 30,300 acres of very scenic river and mountain terrain. The Gunnison River in the park is so "wild" that rafting is not possible. There are many miles of hiking that allow you to explore the majestic beauties of the park. The first number of its zip code is 51 more than the third number. The sum of the first and third numbers is 111. The second number is the first prime number. What is the zip code?
   (You might want to set up two equations for the first and third numbers. They would be *first* + *third* = 111 and *first* - *third* = 51. With a little figuring or guess-and-check, you should get 81 for the first number and 30 for the third number. The second number is 2. Therefore, the zip code is 81230.)

STUDENT PAGE: QUARTER 4: WEEK 6

The theme for week 6 is famous national parks and their zip codes. We will think of a zip code as being composed of a two-digit number, followed by a one-digit number, followed by another two-digit number, e.g., 63435 would be 63, 4, and 35. All of the #2 problems are based on this theme.

Monday   Name: _____

1. **Mental Math:** Perform the following multiplication: _____
   195.38 x 10,000.

2. **National Parks:** Carlsbad Caverns is in New   _____
   Mexico. It contains the deepest limestone cave
   in America, going down to a depth of 1,567 feet.
   The sum of the first and third numbers in the zip
   code is 108, and their difference, first minus third,
   is 68. The second number is the first prime number. What is the zip code?

Tuesday   Name: _____

1. **Calculator Math:** Calculate the following: $\left(1 + \dfrac{0.06}{365}\right)^{5(365)}$ _____

2. **National Parks:** In Alaska, Denali National Park is home to Mount   _____
   McKinley, the highest peak in North America. The first number in
   the zip code has factors of 9 and 11, and the
   third number has factors of 5 and 11. The
   second number is the average of 5
   and 9. What is the zip code?

Wednesday   Name: _____

1. **Estimation:** Estimate the following: 953 / 18.   _____

2. **National Parks:** Volcanoes National Park in Hawaii spans some   _____
   217,000 acres going from sea level to an eleva-
   tion of 13,677 feet at the summit of Mauna Loa.
   This park contains Kilauea, the world's most active volcano. The first number in the zip code has
   five factors of 2 and one factor of 3. The third
   number has two factors of 3 and one factor of 2.
   The second number is the biggest prime less than
   10. What is the zip code?

STUDENT PAGE: QUARTER 4: WEEK 6

Student

Thursday

Name: _____

1. **Calculator Math:** Perform the following addition: _____
   $129.75 + $580.54 + $329.40 + $814.28 + $492.46.

2. **National Parks:** Western Colorado is home to the Dinosaurs National Monument. In 1909 Earl Douglass discovered the fos- _____ sils of dinosaurs at this site. Today there are over 1,500 fossil bones in this park. The first number in the zip code is the square of nine, and the third number is one more than nine. The second number is the first perfect number. What is the zip code?

Friday

Name: _____

1. **Mental Math:** Perform the following addition: 15.78 + 21.97. _____

2. **National Parks:** Big Bend National Park is 801,000 acres in a remote part of west Texas where the Rio Grande River makes a "big turn." The nearest _____ town is over 100 miles away. The first num- ber in the zip code is the biggest prime less than 80. The second is the cube of two, and the third number is an even multiple of 17 that is less than 50. What is the zip code?

Challenge Problem

Name: _____

1. **National Parks:** Black Canyon of the Gunnison National Park is _____ located in southwestern Colorado along the Gunnison River. The park covers some 30,300 acres of very scenic river and moun- tain terrain. The Gunnison River in the park is so "wild" that rafting is not possible. There are many miles of hiking that allow you to explore the ma- jestic beauties of the park. The first number of its zip code is 51 more than the third number. The sum of the first and third numbers is 111. The second number is the first prime number. What is the zip code?

TEACHER PAGE: QUARTER 4: WEEK 7

The theme for week 7 will be making change. You will be given the amount of a purchase and the amount of money given to pay for the purchase. You are to figure the change to give back by using the least amount of bills and coins in your cash register. The register contains the usual coins (pennies, nickels, dimes, and quarters) and the usual bills (ones, fives, tens, and twenties). All of the #2 problems are based on this theme.

Monday

1. **Mental Math:** Perform the following division: $\dfrac{4800}{120}$.

    (You might mention that if you multiply the top by a number, then the answer is multiplied by that number also. Likewise, if you multiply both the top and bottom by the same number, then the answer is the same. In this case you start with $\frac{48}{12} = 4$, next you have $\frac{480}{12} = 40$, and finally you get $\frac{4800}{120} = 40$.)

2. **Make Change:** The purchase is $0.32, and the amount tendered is $1.00.
    (You might suggest that your students try making change mentally by adding up to the amount tendered. Then they could check it with pencil-and-paper methods. The answer is 3-P, 1-N, 1-D, and 2-Q.)

Tuesday

1. **Calculator Math:** Perform the following calculation: $(\sqrt{5} - \sqrt{2}) \times (\sqrt{5} + \sqrt{2})$.
    (You might want to review how to use parentheses and the square root function on your calculator. The answer here is 3. You can perform the following keystrokes: (√ 5 - √ 2) x (√ 5 + √ 2) =. This might lead to some interesting discussion if you wish.)

2. **Make Change:** The purchase is $2.78, and the amount tendered is $5.00.
    (You might suggest that your students try making change mentally by adding up to the amount tendered. Then they could check it with pencil-and-paper methods. The answer is 2-P, 2-D, and 2-$1.)

Wednesday

1. **Estimation:** Estimate the distance you will travel if you average 67 miles per hour for 11 hours 30 minutes.
    (You might want to review Distance = Rate x Time and tell your students to round to make the estimate. In this case you would get 70 x 10 = 700 miles.)

2. **Make Change:** The purchase is $3.17, and the amount tendered is $5.02.
    (You might suggest that your students try making change mentally by adding up to the amount tendered. Then they could check it with pencil-and-paper methods. The answer is 1-D, 3-Q, and 1-$1.)

TEACHER PAGE: QUARTER 4: WEEK 7

Thursday

1. **Calculator Math:** Calculate the distance you will travel if you average 67 miles per hour for 11 hours 30 minutes.
   (You might want to review Distance = Rate x Time and tell your students to convert the time into the single unit of hours. You can convert 11 hours 30 minutes to 11 + 30 / 60 = 11.5 hours, and then perform the following keystrokes: 67 x 11.5 =. The answer is 67 miles x 11.5 hours = 770.5 miles.)

2. **Make Change:** The purchase is $7.17, and the amount tendered is $12.25.
   (You might suggest that your students try making change mentally by adding up to the amount tendered. Then they could check it with pencil-and-paper methods. The answer is 3-P, 1-N, and 1-$5.)

Friday

1. **Mental Math:** You buy 20 pieces of candy for $0.35 each, and you sell them for $0.60 each. How much money do you make?
   (You might tell them to figure out the profit per piece first and then multiply that by 20. They should be able to do this mentally and get $5.00 as the answer.)

2. **Make Change:** The purchase is $19.03, and the amount tendered is $50.
   (You might suggest that your students try making change mentally by adding up to the amount tendered. Then they could check it with pencil-and-paper methods. The answer is 2-P, 2-D, 3-Q, 1-$10, and 1-$20.)

Challenge Problems

1. **Make Change:** The purchase is $71.11, and the amount tendered is $100.15.
   (You might suggest that your students try making change mentally by adding up to the amount tendered. Then they could check it with pencil-and-paper methods. The answer is 4-P, 4-$1, 1-$5, and 1-$20.)

2. **Make Change:** The purchase is $113.43, and the amount tendered is $200.
   (You might suggest that your students try making change mentally by adding up to the amount tendered. Then they could check it with pencil-and-paper methods. The answer is 2-P, 1-N, 2-Q, 1-$1, 1-$5, and 4-$20.)

3. **Make Change:** The purchase is $0.79, and the amount tendered is $20.
   (You might suggest that your students try making change mentally by adding up to the amount tendered. Then they could check it with pencil-and-paper methods. The answer is 1-P, 2-D, 4-$1, 1-$5, and 1-$10.)

STUDENT PAGE: QUARTER 4: WEEK 7

Student

The theme for week 7 will be making change. You will be given the amount of a purchase and the amount of money given to pay for the purchase. You are to figure the change to give back by using the least amount of bills and coins in your cash register. The register contains the usual coins (pennies, nickels, dimes, and quarters) and the usual bills (ones, fives, tens, and twenties). All of the #2 problems are based on this theme.

Monday　　　　　　　　　　　　　　Name:_____

1. **Mental Math:** Perform the following division: $\frac{4800}{120}$.　　_____

2. **Make Change:** The purchase is $0.32, and the amount tendered is $1.00.　　_____

Tuesday　　　　　　　　　　　　　　Name:_____

1. **Calculator Math:** Perform the following calculation: $(\sqrt{5}-\sqrt{2}) \times (\sqrt{5}+\sqrt{2})$.　　_____

2. **Make Change:** The purchase is $2.78, and the amount tendered is $5.00.　　_____

Wednesday　　　　　　　　　　　　　Name:_____

1. **Estimation:** Estimate the distance you will travel if you average 67 miles per hour for 11 hours 30 minutes.　　_____

2. **Make Change:** The purchase is $3.17, and the amount tendered is $5.02.　　_____

STUDENT PAGE: QUARTER 4: WEEK 7

Student

Thursday　　　　　　　　　　　　　　　　　Name:_____

1. **Calculator Math:** Calculate the distance you will travel if you _____
average 67 miles per hour for 11 hours 30 minutes.

2. **Make Change:** The purchase is $7.17, and the _____
amount tendered is $12.25.

Friday　　　　　　　　　　　　　　　　　　Name:_____

1. **Mental Math:** You buy 20 pieces of candy for $0.35 each, and _____
you sell them for $0.60 each. How much money do you make?

2. **Make Change:** The purchase is $19.03, and the _____
amount tendered is $50.

Challenge Problems　　　　　　　　　　　　Name:_____

1. **Make Change:** The purchase is $71.11, and the amount ten- _____
dered is $100.15.

2. **Make Change:** The purchase is $113.43, and the amount ten- _____
dered is $200.

3. **Make Change:** The purchase is $0.79, and the _____
amount tendered is $20.

　　　　　　　136

TEACHER PAGE: QUARTER 4: WEEK 8    Teacher

The theme for week 8 is triangle, rectangle, and circle geometry. You will be given certain important dimensions of a shape that is composed of triangles, rectangles, and circles, and you will be asked to find the area of the shape. All of the #2 problems are based on this theme.

Monday

1. **Mental Math:** Calculate 75% of 28.
   (You might want to mention that it would be helpful to think of 75% as $\frac{3}{4}$. Therefore, you are trying to find $\frac{3}{4}$ of 28 = 3 x ($\frac{1}{4}$ x 28) = 3 x 7 = 21.)

2. **T-R-C Geometry:** Find the area of the whole region shown.

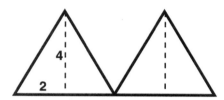

   (You might want to review the area of triangles and have your students look at the whole as the sum of its components. Each triangle has area = $\frac{1}{2}$ x 4 x 4 = 8. The answer is 16.)

Tuesday

1. **Calculator Math:** Calculate: $\left(1 + \dfrac{0.0425}{12}\right)^{12 \times 2}$.
   (You might want to show your students how to use the exponential key on their calculators. Also, you might want to tell them to do the operation inside the parentheses starting with the division. You can perform the following keystrokes: .0425 / 12 = + 1 = ^ (12 x 2) =. The answer is 1.08855.)

2. **T-R-C Geometry:** Find the area of the whole region shown.

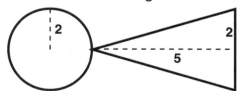

   (You might want to review the area of triangles and circles and have your students look at the whole as the sum of its components. The area of the circle is r^2 x π = 2 x 2 x π = 4π, and the area of the triangle is $\frac{1}{2}$ x 4 x 5 = 10. Thus the total area is 10 + 4π = 22.57.)

Wednesday

1. **Estimation:** You work three days a week at a part-time job from 9:30 a.m. to 1:00 p.m. If you earn $6.25 an hour, then estimate how much you make each week.
   (You might want to help your students figure that from 9:30 a.m. to 1:00 p.m. is $3\frac{1}{2}$ hours. Therefore, the person works approximately 10 hours a week and so makes about $62.50 each week.)

TEACHER PAGE: QUARTER 4: WEEK 8   Teacher

Wednesday (continued)

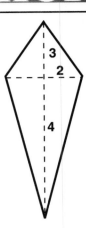

2. **T-R-C Geometry:** Find the area of the whole region shown.
(You might want to review the area of triangles and have your students look at the whole as the sum of its components. The whole is made up of four triangles. The upper two each have area $= \frac{1}{2}$ x 2 x 3 = 3, and the lower two each have area $= \frac{1}{2}$ x 2 x 4 = 4. Therefore, the whole has area = 3 + 3 + 4 + 4 = 14. You could also look at it as two triangles. The upper triangle having area $= \frac{1}{2}$ x 4 x 3 = 6, and the lower triangle having area $= \frac{1}{2}$ x 4 x 4 = 8. The total area = 6 + 8 = 14.)

Thursday

1. **Calculator Math:** You work three days a week at a part-time job from 9:30 a.m. to 1:00 p.m. If you earn $6.25 an hour, then calculate how much you make each week.
(You might want to help your students figure that from 9:30 a.m. to 1:00 p.m. is $3\frac{1}{2}$ hours. Therefore, the answers will be 3 x 3.5 x $6.25 = $65.63.)

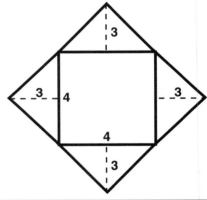

2. **T-R-C Geometry:** Find the area of the whole region shown.
(You might want to review the area of triangles and rectangles and have your students look at the whole as the sum of its components. The whole is the sum of four identical triangles plus one square. Each triangle has area $= \frac{1}{2}$ x 4 x 3 = 6, and the square has area = 4 x 4 = 16. The answer is 40.)

Friday

1. **Mental Math:** Add the following numbers: 57 + 84 + 62.
(You might want to tell your students that these three numbers can be added mentally even without compatible numbers. Just add 50 + 80 + 60 = 190 and 7 + 4 + 2 = 13; hence you get 190 + 13 = 203.)

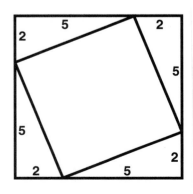

2. **T-R-C Geometry:** Find the area of the square inside the whole region shown.
(You might want to review the area of triangles and rectangles. You also might want to stress that the whole is the sum of its components. In this case, the whole is clearly a square of area = 7 x 7 = 49. The components are four identical triangles and one square. The area of each triangle is $\frac{1}{2}$ x 2 x 5 = 5, so the total area of the triangles is 20. Therefore, the area of the inside square is 49 - 20 = 29.)

STUDENT PAGE: QUARTER 4: WEEK 8

The theme for week 8 is triangle, rectangle, and circle geometry. You will be given certain important dimensions of a shape that is composed of triangles, rectangles, and circles, and you will be asked to find the area of the shape. All of the #2 problems are based on this theme.

- -

Monday   Name: _____

1. **Mental Math:** Calculate 75% of 28.

2. **T-R-C Geometry:** Find the area of the whole region shown.

- -

Tuesday   Name: _____

1. **Calculator Math:** Calculate: $\left(1 + \dfrac{0.0425}{12}\right)^{12 \times 2}$.

2. **T-R-C Geometry:** Find the area of the whole region shown.

- -

Wednesday   Name: _____

1. **Estimation:** You work three days a week at a part-time job from 9:30 a.m. to 1:00 p.m. If you earn $6.25 an hour, then estimate how much you make each week.

2. **T-R-C Geometry:** Find the area of the whole region shown.

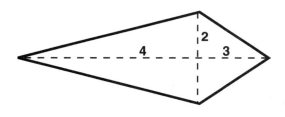

STUDENT PAGE: QUARTER 4: WEEK 8

Student

Thursday

Name: _____

1. **Calculator Math:** You work three days a week at a part-time job _____
   from 9:30 a.m. to 1:00 p.m. If you earn $6.25 an hour, then cal-
   culate how much you make each week.

2. **T-R-C Geometry:** Find the area of the whole region shown. _____

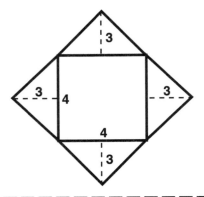

Friday

Name: _____

1. **Mental Math:** Add the following numbers: 57 + 84 + 62. _____

2. **T-R-C Geometry:** Find the area of the square inside the whole _____
   region shown.

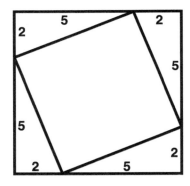

TEACHER PAGE: QUARTER 4: WEEK 9

The theme for week 9 is guess that shape. All of the #2 problems are based on this theme. For each of these problems, you will be given a set of clues. You will then have to become a mathematical detective and try to figure out the most descriptive name for the plane polygonal shape.

Monday

1. **Mental Math:** Find $66\frac{2}{3}\%$ of 45.

   (You might mention that, if you can think of the percent as $\frac{2}{3}$, you can calculate $\frac{2}{3} \times 45 = 30$.)

2. **Guess That Shape:** This shape has four sides. All four sides and angles are equal. Guess that shape.
   (You might want to review naming four-sided polygons. Be careful in this problem. The answer is a square.)

Tuesday

1. **Calculator Math:** Perform the calculation: $\dfrac{2}{\frac{2}{3} + \frac{3}{4}}$.

   (You might want to talk about the importance of order in this calculation. You might want your students to use the reciprocal key on their calculators. As usual, you should tell your students that they shouldn't use memory or write down intermediate answers. You can perform the following keystrokes: $2 / 3 = + 3 / 4 = x^{-1} =$. The answer is $\frac{24}{17} = 1.41176$ or $1\frac{7}{17}$.)

2. **Guess That Shape:** This shape has four sides. Opposite sides of this shape resemble railroad tracks, i.e., they are parallel. However, we don't know anything about the lengths of the different opposite pairs. Guess that shape.
   (You might want to review naming four-sided polygons. The answer is a parallelogram.)

Wednesday

1. **Estimation:** If four bagels cost $1.77, then estimate how much a dozen bagels will cost.
   (You might want to have your students figure how many 4's are in a dozen and round the price to $1.80. Thus you get approximately $5.40.)

2. **Guess That Shape:** This shape has four sides. One pair of opposite sides are parallel and the other is not, but those two sides are equal in length. Guess that shape.
   (You might want to review naming four-sided polygons. The answer is an isosceles trapezoid.)

TEACHER PAGE: QUARTER 4: WEEK 9

Thursday

1. **Calculator Math:** If four bagels cost $1.77, then how much does a dozen bagels cost? (You might want to have your students figure out how many fours are in a dozen and then simply multiply that number by $1.77. Since there are 12 / 4 = 3 fours in 12, then you can get the cost by performing the following keystrokes: 3 x 1.77 =. The answer is $5.31.)

2. **Guess That Shape:** This shape has four sides. They are all equal in length. However, the shape doesn't necessarily have any right angles. Guess that shape. (You might want to review naming four-sided polygons. The answer is a rhombus.)

Friday

1. **Mental Math:** Add the following numbers: 4.88 + 8.93. (You might want to mention that if you add a quantity to one number and subtract it from the other, the sum will remain the same. In our problem, add 0.07 to the right number and subtract it from the left number giving 4.81 + 9.00 = 13.81.)

2. **Guess That Shape:** This shape has eight sides. All the sides are equal in length, and all the angles are equal in measure. You will often see models of this shape on the street. Guess that shape. (You might want to review naming eight-sided polygons. The answer is a regular octagon. Note! *Regular* in the name means that all the sides are equal.)

Challenge Problems

1. **Guess That Shape:** This shape is called a heptagon. It has one more side than a hexagon. Guess how many sides this shape has. (This may be a difficult name for the student to remember, but the clue should guide them to the correct answer of 7 sides.)

2. **Guess That Shape:** This shape is called a decagon. It has twice the number of sides as that of a pentagon. Guess how many sides this shape has. (This may be a difficult name for the student to remember, but the clue should guide them to the correct answer of 10 sides.)

STUDENT PAGE: QUARTER 4: WEEK 9

Student

The theme for week 9 is guess that shape. All of the #2 problems are based on this theme. For each of these problems, you will be given a set of clues. You will then have to become a mathematical detective and try to figure out the most descriptive name for the plane polygonal shape.

Monday   Name: _____

1. **Mental Math:** Find $66\frac{2}{3}\%$ of 45.   _____

2. **Guess That Shape:** This shape has four sides. All four sides and angles are equal. Guess that shape.

   _____

Tuesday   Name: _____

1. **Calculator Math:** Perform the calculation: $\dfrac{2}{\frac{2}{3}+\frac{3}{4}}$.   _____

2. **Guess That Shape:** This shape has four sides. Opposite sides of this shape resemble railroad tracks, i.e., they are parallel. However, I don't know anything about the lengths of the different opposite pairs. Guess that shape.

   _____

Wednesday   Name: _____

1. **Estimation:** If four bagels cost $1.77, then estimate how much a dozen bagels will cost.   _____

2. **Guess That Shape:** This shape has four sides. One pair of opposite sides are parallel and the other is not, but those two sides are equal in length. Guess that shape.

   _____

143

STUDENT PAGE: QUARTER 4: WEEK 9   Student

Thursday

Name:_____

1. **Calculator Math:** If four bagels cost $1.77, then how much does _____
   a dozen bagels cost?

2. **Guess That Shape:** This shape has four
   sides. They are all equal in length. However, the shape doesn't necessarily have
   any right angles. Guess that shape. _____

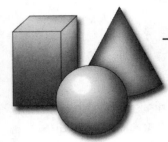

Friday

Name:_____

1. **Mental Math:** Add the following numbers: 4.88 + 8.93. _____

2. **Guess That Shape:** This shape has eight
   sides. All the sides are equal in length, and
   all the angles are equal in measure. You
   will often see models of this shape on the
   street. Guess that shape. _____

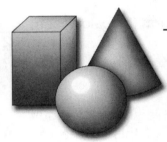

Challenge Problems

Name:_____

1. **Guess That Shape:** This shape is called a heptagon. It has one _____
   more side than a hexagon. Guess how many sides this shape
   has.

2. **Guess That Shape:** This shape is called a
   decagon. It has twice the number of sides _____
   as that of a pentagon. Guess how many
   sides this shape has.

Glossary of Terms

Acute angle: An angle between 0 and 90 degrees in measure

Acute triangle: A triangle whose three angles are all acute

Average (or Mean): The average (or mean) of a set of numbers is the number calculated by summing all the numbers in the set and dividing this sum by the number of numbers in the set.

Circumference: The distance around the circle; it is a linear measure.

Composite number: A counting number with more than two factors

Counting numbers: The numbers 1, 2, 3, 4, 5, 6, …

Cube root: The number b such that $b \times b \times b = a$ (e.g., 2 is the cube root of 8 because $2 \times 2 \times 2 = 8$.)

Diameter: The distance from one side of a circle to the other through the center

Divisor: A counting number b is a divisor of an integer a if the counting number b is a factor of the integer a (e.g., 4 is a divisor of 12 because 4 is a factor of 12.)

Equilateral triangle: A triangle where all three sides are equal in length

Factor: A counting number b is a factor of an integer a if there is another integer c such that a is b times c (e.g., 4 is a factor of 12 since 12 = 4 × 3.)

Hexagon: A polygon with exactly six sides

Integers: The set of whole numbers plus the negatives of the counting numbers (e.g., the numbers … -3, -2, -1, 0, 1, 2, 3, …)

Isosceles trapezoid: A trapezoid whose nonparallel sides are equal in length

Isosceles triangle: A triangle with two sides that are equal in length

Kite: A quadrilateral with two non-overlapping pairs of adjacent equal sides

Measure of an angle: The least number of degrees (or radians) of revolution from one side to the other side of an angle

Obtuse angle: An angle that has measure between 180 to 360 degrees

Obtuse triangle: A triangle with exactly one angle that is obtuse

Octagon: A polygon with exactly eight sides

Parallelogram: A quadrilateral with both pairs of opposite sides that are parallel

Pentagon: A polygon with exactly five sides

Perimeter: The distance around a polygon

Polygon: A figure that can be traced using straight line segments that start and end at the same point and do not cross any point except the starting/ending point

Prime number: A counting number that has exactly two factors

Quadrilateral: A polygon with exactly four sides

Radius: The distance from the center of a circle to the circumference of the circle

Glossary of Terms

Rhombus: A quadrilateral with all four sides equal

Right angle: An angle that has a measure of 90 degrees

Right triangle: A triangle that has one right angle

Scalene triangle: A triangle that has no two sides of equal length

Square: A quadrilateral with all four sides equal and all four angles are right angles

Square root: The square root of a number a is the number b that multiplied by itself gives a (e.g., 3 is the square root of 9 since $9 = 3 \times 3$.)

Straight angle: An angle that has a measure of 180 degrees

Trapezoid: A quadrilateral that has exactly one pair of parallel sides

Whole numbers: The counting numbers together with 0 (e.g., the numbers 0, 1, 2, 3, 4, ...)